The great groups of fish

the yellow areas show how much the groups expanded during the various periods

THE HISTORY OF LIFE ON EARTH

MARINE LIFE

© 1986 English-language edition by Facts On File, Inc.
460 Park Avenue South, New York, NY 10016

© 1986 Editoriale Jaca Book Spa, Milano

All rights reserved. No part of this book may be reproduced or utilized in any form or by any means, electronic or mechanical, including photocopying, recording or by any information storage and retrieval systems, without permission in writing from the Publisher.

editorial coordination
CATERINA LONGANESI

CONTENTS

1. Our earliest origins
2. Extinct *Agnatha*: the ostracoderms
3. Existing *Agnatha*: the myxinoid cyclostomes
4. Existing *Agnatha*: the petromyzon cyclostomes
5. How a mouth for preying was invented
6. The first vertebrates with real mouths: the placoderms
7. The Devonian Period: the great age of fish
8. How cartilaginous and bony fish solved the problem of osmotic pressure
9. How the two types of fish solved the problem of specific weight
10. Fish at the end of the Devonian Period
11. Cartilaginous fish, or the chondrichthyes
12. Migration of the mouth and sense of smell in the shark
13. The magnificent sense organs of the shark: the eye and the lateral line
14. What sharks eat and how
15. How sharks swim and hunt
16. How sharks reproduce
17. The survivors: living sharks
18. Rayfish, or batoids
19. The electrical organs of certain batoids
20. Bony fish: the chondrosts
21. The holosts
22. Modern fish: the teleosts
23. The shape and color of the teleosts
24. The teleosts: how they hunt and why they emigrate
25. The teleosts: a detail or two
26. Reproduction and parental care in the teleosts
27. The teleosts of the abyss
28. Sarcopterygii: the fish that also breathe air

Library of Congress Card Catalog no. 86-81886

ISBN 0-8160-1556-2

color separation by
Carlo Scotti, Milano
photosetting by
Elle Due, Milano
printed and bound in Italy by
Tipolitografia G. Canale & C. Spa, Torino

MARINE LIFE

Giuseppe Minelli
Professor of Comparative Anatomy
University of Bologna, Italy

illustrated by the
Lorenzo Orlandi Studio

translated and adapted by
Bryan Fleming

the "History of Life on Earth" series
is conceived, designed and produced by
Jaca Book

Facts On File Publications
New York, New York • Oxford, England

In the seas of the Paleozoic era, more than 600 million years ago, there were invertebrates with the strangest shapes. The *Pikaia* was one of the few that were ancestors of the vertebrates.

1. OUR EARLIEST ORIGINS

In the Earth's waters in the Precambrian-Cambrian Period, over 600 million years ago, there were countless forms of life, an incredible variety. We are fortunate, by some lucky chance, to still have imprints of the animals of that period, even though they did not have hard parts, like shells, which fossilize easily.

THE FIRST ANIMALS

Some of these animals can be identified with the great groups described by zoology, such as the *Wiwaxia*, with the mollusks, and the *Peytoia*, shaped like a slice of pineapple, a forerunner of the coelenterates. Others were unknown and, to our eyes today, utterly absurd: the *Hallucigenia*, with lots of little mouths and 12 stilt-like legs, and the *Opabina*, with a large mouth on the end of a long tentacle.

TWO PATHS OF EVOLUTION

But what was going on in those waters so long ago has a lot to do with us today. Animal-life was evolving along two different paths. On one path were the invertebrates, with their own special mode of development and anatomical organization. On the other were the chordates, rope-like creatures that evolved into the vertebrates — and fish, the subject of this volume, are vertebrates. It is curious to note that the force of evolution, the ability to create new forms, was completely different along our two paths. Apart from a number of curiously shaped creatures, all inevitably destined to disappear from the face of the Earth, the invertebrates spread rapidly, developing forms that were very refined as movers, hunters and defenders, while the chordates took a long time to make their mark. They may already have existed in the seas of the Cambrian Period, in the shape of little swimming animals like the *Pikaia*, but they were a minority compared to the invertebrates. There are only a few living representatives of these ancient fish forerunners left, and they probably come from the sterile branches of the chordates. They include the urochords, or tunicates; like the ascidians, the actual cord, called the notocord, is only in the tail and only in the larval stage, because the adult, with its barrel-shaped mouth, has nothing whatever to do with the general organization of the chordates.

THE *AMPHIOXUS*

The most interesting was the *Amphioxus*, or lancelet. It was small, blind, practically devoid of sense organs and an abysmal swimmer. This little creature,

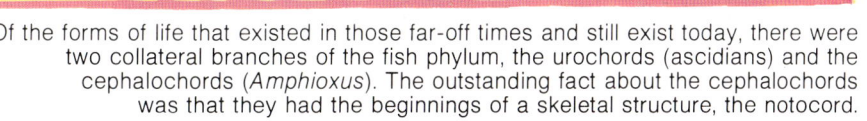

Of the forms of life that existed in those far-off times and still exist today, there were two collateral branches of the fish phylum, the urochords (ascidians) and the cephalochords (*Amphioxus*). The outstanding fact about the cephalochords was that they had the beginnings of a skeletal structure, the notocord.

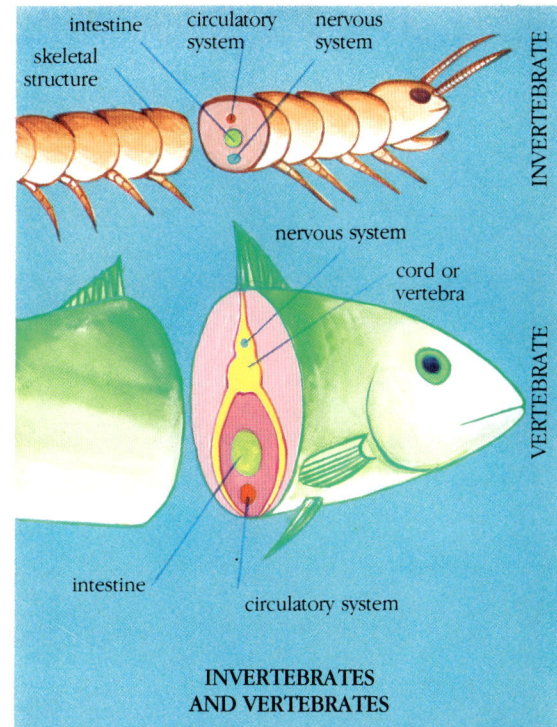

INVERTEBRATES AND VERTEBRATES

Very early on in the history of life on Earth, there was a very clear subdivision of multicellular animals into two large groups that differed in the layout of certain fundamental organs.

The rigid skeletal structures of the *invertebrates* were on the outside of the body and their nervous system was "ventral" to the intestine, below it, while the heart was "dorsal" to the intestine, above it.

The exact opposite was the case with the *vertebrates*: the rigid skeletal structures, cartilaginous or bony, were inside the body, and the heart was ventral.

These two large groups that separated in the Earth's waters in Precambrian times subsequently developed quite independently, the invertebrates faster and earlier, the chordates and vertebrates more slowly and later, but their results were far more spectacular.

still with us today, is considered to be an ancient representative of the fish, a very ancient one.

A lot of special research has gone into finding out everything possible about the shy creature. In truth, it has been estimated that after man and the mouse, the amphioxus is the most closely studied and best known of all animals.

This precious little inhabitant of the seas, just 2 to 2½ inches long, usually lives well-dug into the sand. The only part that emerges is the mouth, with all the tendrils that help it filter water. All the precursors, and the first vertebrates too, all had this same characteristic. It is, in fact, a very primitive way of feeding: mere suction and filtration of water, to get the little bits of food suspended there. It will be a long time before we see a more efficient feeding system evolve.

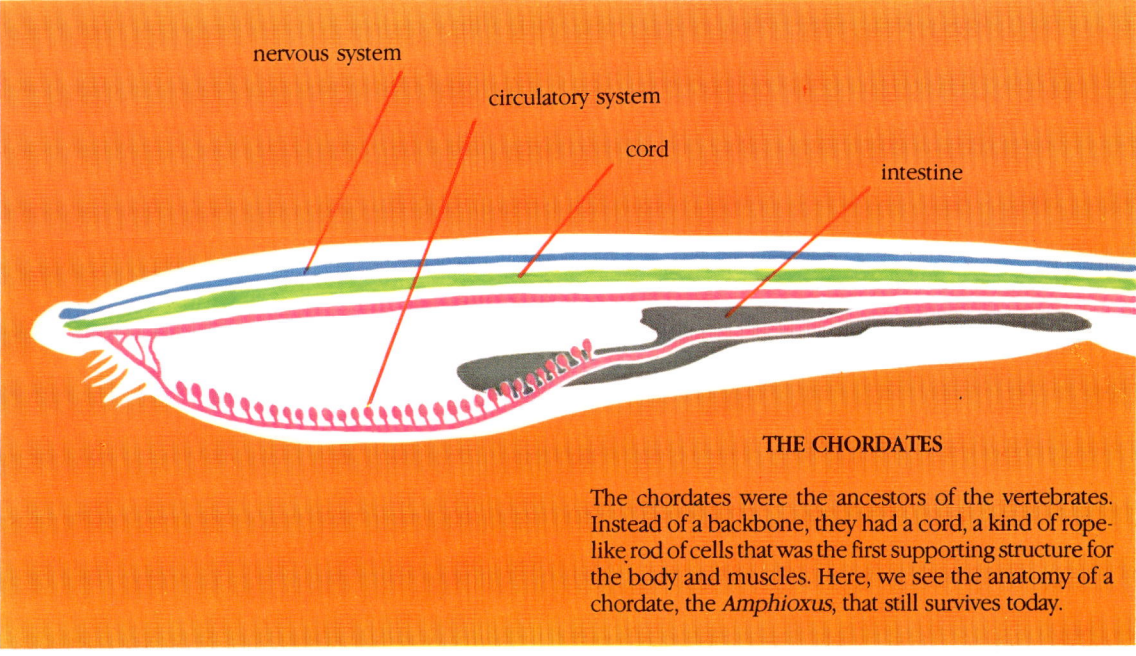

THE CHORDATES

The chordates were the ancestors of the vertebrates. Instead of a backbone, they had a cord, a kind of rope-like rod of cells that was the first supporting structure for the body and muscles. Here, we see the anatomy of a chordate, the *Amphioxus*, that still survives today.

In the waters of the Silurian Period, 450 million years ago, the first vertebrates appeared. They were the ostracoderms, and they dominated the Earth's waters for 60 million years before becoming extinct.

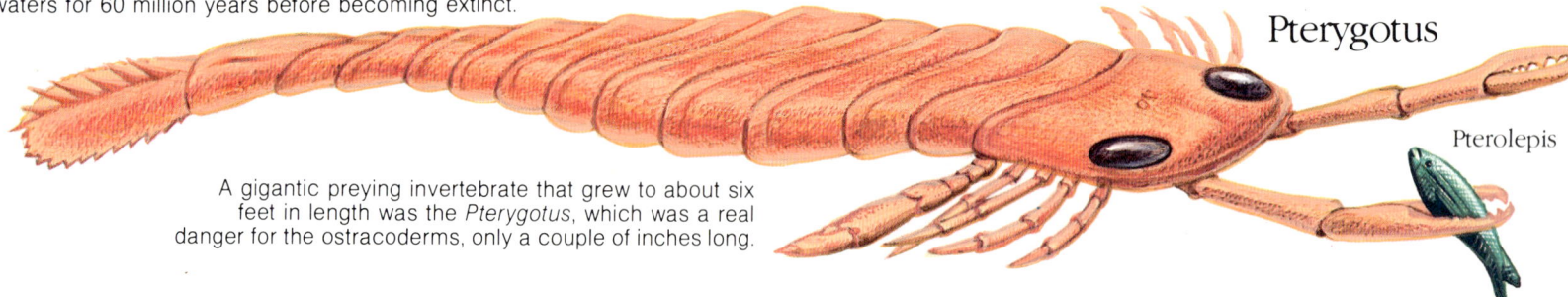

A gigantic preying invertebrate that grew to about six feet in length was the *Pterygotus*, which was a real danger for the ostracoderms, only a couple of inches long.

2. EXTINCT AGNATHA: THE OSTRACODERMS

As soon as the first vertebrates appeared on the face of the Earth, a little mystery set in. The probable forerunners of fish had been small animals with no skeletal structure, slender, transparent, and practically unable to fossilize.

THE FIRST VERTEBRATES

And then, quite suddenly, 400 to 500 million years ago, in the waters of the Ordovician Era, the first real vertebrates made their appearance. They were the ostracoderms (shell-skins), still very small, a few inches long, but armored with the heavy, bony structure covering their heads and the front part of their bodies that gave them their name.

The mystery is how this armor-plating came to be. There is always some logic, some good reason behind the structures and organization of anatomy. One hypothesis says that their shield-like covering was to protect them against the high concentration of salt in the sea (the earliest ostracoderms did, in fact, inhabit the seas) or against a precipitation of calcium salts deriving from their excessive concentration in the seas of the period. But other animals living in the seas at the same time may well have been the real reason; these were the arthropods (with limbs), including the *Pterygotus*, the old ancestors of lobsters and shrimps, gigantic at the time, about six feet long. These hunter fish, with their strong nippers and claws, were probably very bitter enemies of the first, defenseless vertebrates to whom robust armor-plating could offer protection. It may well have been the fight for survival against those first gigantic arthropods that caused the ostracoderms to form their bony shield.

THE UNMOVABLE MOUTH

Although they are extinct, we do know about the soft parts of the ostracoderms — brain, heart, gills, intestine, sense organs — because they left imprints on the inner surface of their armor-plating, and this fossilized perfectly. Study of these soft parts has revealed that these creatures of almost half a billion years ago already had the typical anatomical organization of the vertebrates, with the same apparatus and sense organs. But their mouth was very primitive, unmovable, circular, usable only for sucking in seawater and filtering the organic substances in it.

This poor form of nourishment placed very severe restrictions on the development of their body. Generally speaking, the ostracoderms never grew much longer than a few inches. But their anatomical organization offset that.

THE EXPANSION OF THE OSTRACODERMS

Evolution soon brought about the appearance of various species of ostracoderm. They invaded not only the deep seas but coastal waters as well, and lakes and rivers, and, for more than 60 million years, an extremely long time, they were the unchallenged

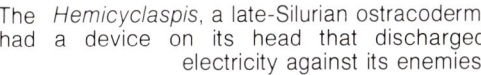

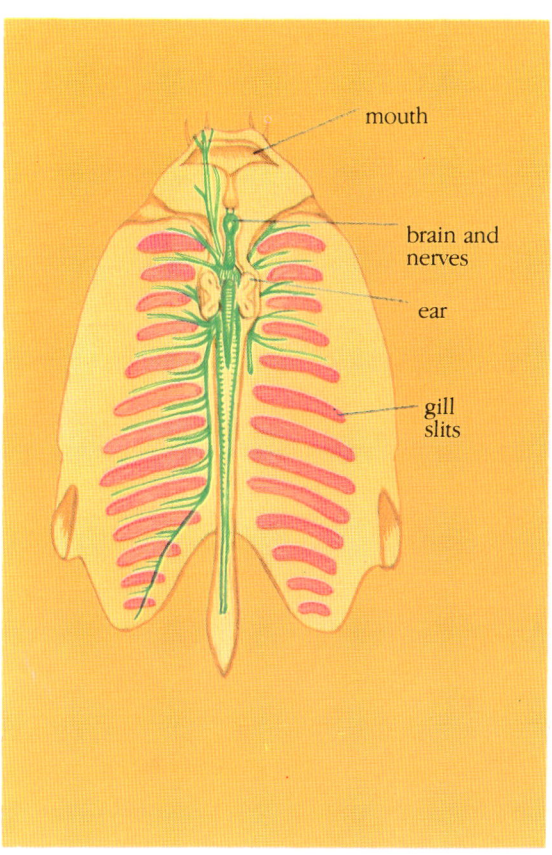

A bony armor-plating developed around the soft structures of the ostracoderms. As it fossilized, the shield maintained the most minute imprints, so we can reconstruct ostracoderm anatomy very accurately. The inner surface of the armor-plating is shown in light yellow in the drawing.

The *Hemicyclaspis*, a late-Silurian ostracoderm, had a device on its head that discharged electricity against its enemies.

THE AGNATHA

The outstanding characteristic that all the first vertebrates that appeared on the Earth had in common was the mouth. In all of them, this fundamental instrument for capturing food was immovable, devoid, that is, of articulated mandibular arches — jaws — that could move. Their mouth was circular and held wide open by pieces of cartilage.

This meant that the animal could not tear off food and chew it. It had to be happy with sucking in liquids and filtering them to trap the organic particles suspended there.

This is the origin of the term *agnatha*, a Greek word that means "jawless" and includes them all: from the now-extinct ostracoderms to the still-surviving cyclostomes.

masters of all the Earth's waters. This was partly due to the gradual disappearance of the gigantic *Pterygoti*, the only creatures that were big, strong and aggressive enough to compete with them. And evolution also softened the stubby, squat structure of these first ostracoderms; it lightened and finally did away with the heavy protective shield of bone.

The *Pterolepis*, for instance, was already quite a smart-looking fellow, something like a real little fish, and he had no shield.

The powerful muscles of his trunk and tail already made him a strong, fairly fast swimmer. But the usual circular, unmovable mouth leads us to believe that this creature too spent his life dredging the bottom of the seas, filtering sand and mud for food.

The saving grace was the asymmetrical tail, with the lower flap longer than the upper. This gave the *Pterolepis* enough thrust to sink his head into the bottom in search of nourishment.

There must have been a multitude of lurking dangers for these little animals. One group of ostracoderms, the cephalaspes, and the *Hemicyclaspis* in particular, sprouted a long, narrow structure on their heads. This has been tentatively identified as a very voluminous electrical organ that gave out shocks to discourage would-be enemies.

DECLINE AND FALL OF THE OSTRACODERMS

The reign of the ostracoderms began its decline toward the beginning of the Devonian Period, with the advent of the first real fish, endowed with upper and lower jaws and, therefore, quite capable of hunting, killing and devouring their prey. The first of these real fish were the placoderms (plated skins), followed by the acanthodians (spiny fish), about whom we shall have more to say later. Slowly, but inevitably, the ostracoderms became extinct, unable as they were to withstand the overwhelming might of the newcomers. They all disappeared, except for a small group that still lives on today, the cyclostomes (round mouths).

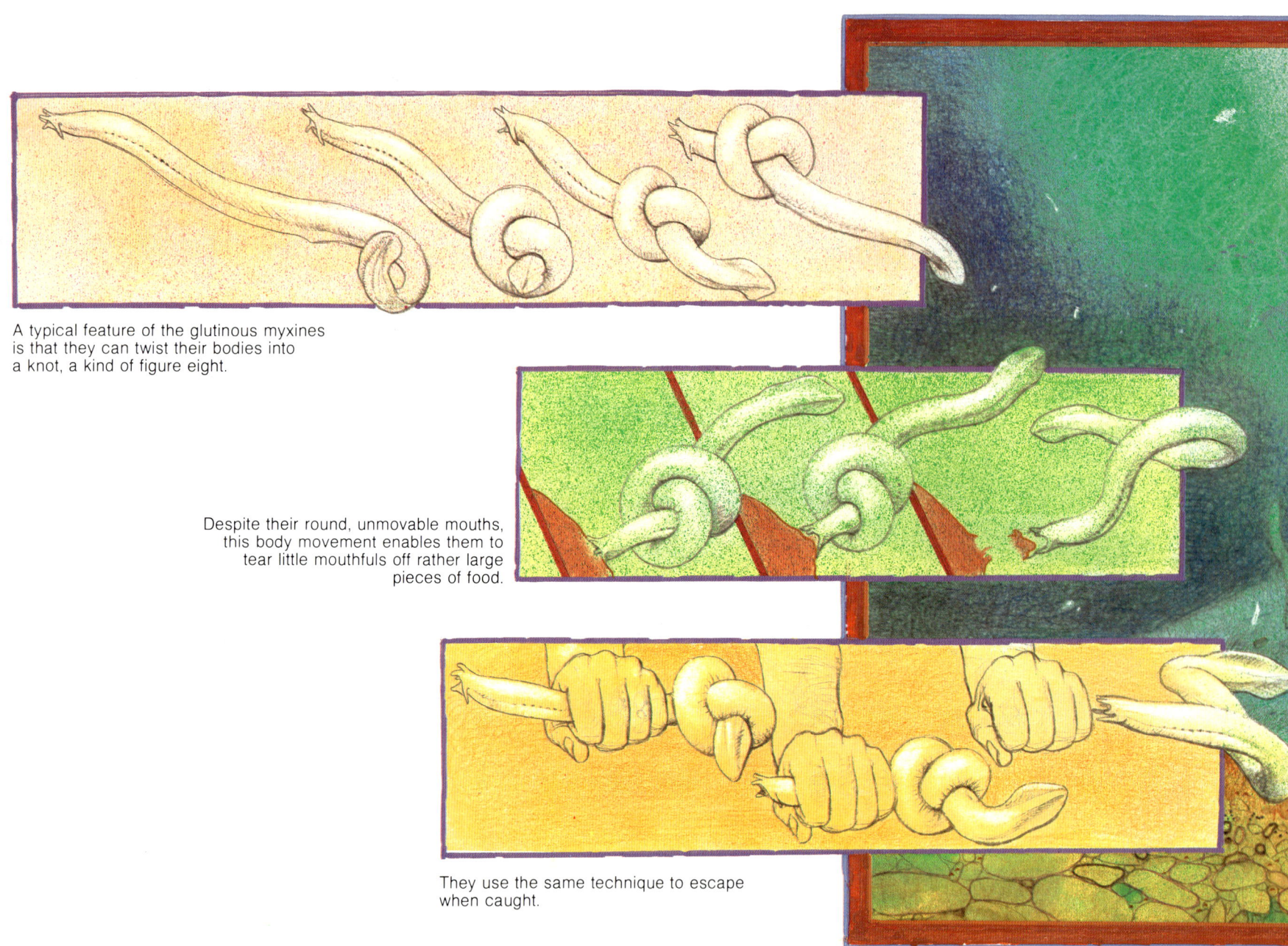

A typical feature of the glutinous myxines is that they can twist their bodies into a knot, a kind of figure eight.

Despite their round, unmovable mouths, this body movement enables them to tear little mouthfuls off rather large pieces of food.

They use the same technique to escape when caught.

3. EXISTING AGNATHA: THE MYXINOID CYCLOSTOMES

Only a few members of the great agnatha empire still live today, in two very different orders of cyclostomes: the myxinoids, or hagfish, and the petromyzons, or lampreys.

THE MYXINES

Hagfish are the most characteristic and best known of the order.

They are about a foot long and have eel-like bodies and a definitely regressive, or backward, anatomy compared to the typical vertebrate. Their sense organs comprise no more than a very refined sense of smell, for their eyes and ears are practically useless.

The reason for this regression, or backwardness, as we might call it, is their extremely monotonous life: all they do is wait for victims to parasitize, or to feed off. Hagfish live on the sea bottom, at very great depths, where they twist and turn in the sand, swimming backwards until half their length is buried in the sand. And there they wait, mouth and single nostril poking out, ready to catch the scent of food.

THE SEA SCAVENGERS

Hagfish are the scavengers of the sea. Nearly everything edible that falls to the bottom is immediately surrounded by dozens of these myxines, who are attracted by the odor and make a real banquet of it. But live fish too, especially if they have some difficulty in moving, are also a feast for these primitive vertebrates. They have a very original way of attacking their prey: they slip in among the gills of the victim and begin to secrete a slimy liquid that covers the respiratory tissue and eventually suffocates the poor fish. Once the wretched victim is dead, the hagfish begins to gobble it up, beginning on the inside and exiting from the fish only when it is no more than an empty frame of skin and scales. The myxines can also be great nuisances to fishermen; they are often quite capable of attacking fish caught up in nets or on hooks on the sea bed and destroying a whole catch in a couple of hours.

HOW HAGFISH DEVOUR THEIR FOOD

The myxines are agnatha, that is, they have unmovable mouths, so they should not really be able to gobble up other fish. But they have adopted a very curious method of eating their dinner. Their mouth is sucker-like and takes a firm hold on the prey. The streamlined body then contracts into a kind of knot that slides up to to the head, detaching it, along with a

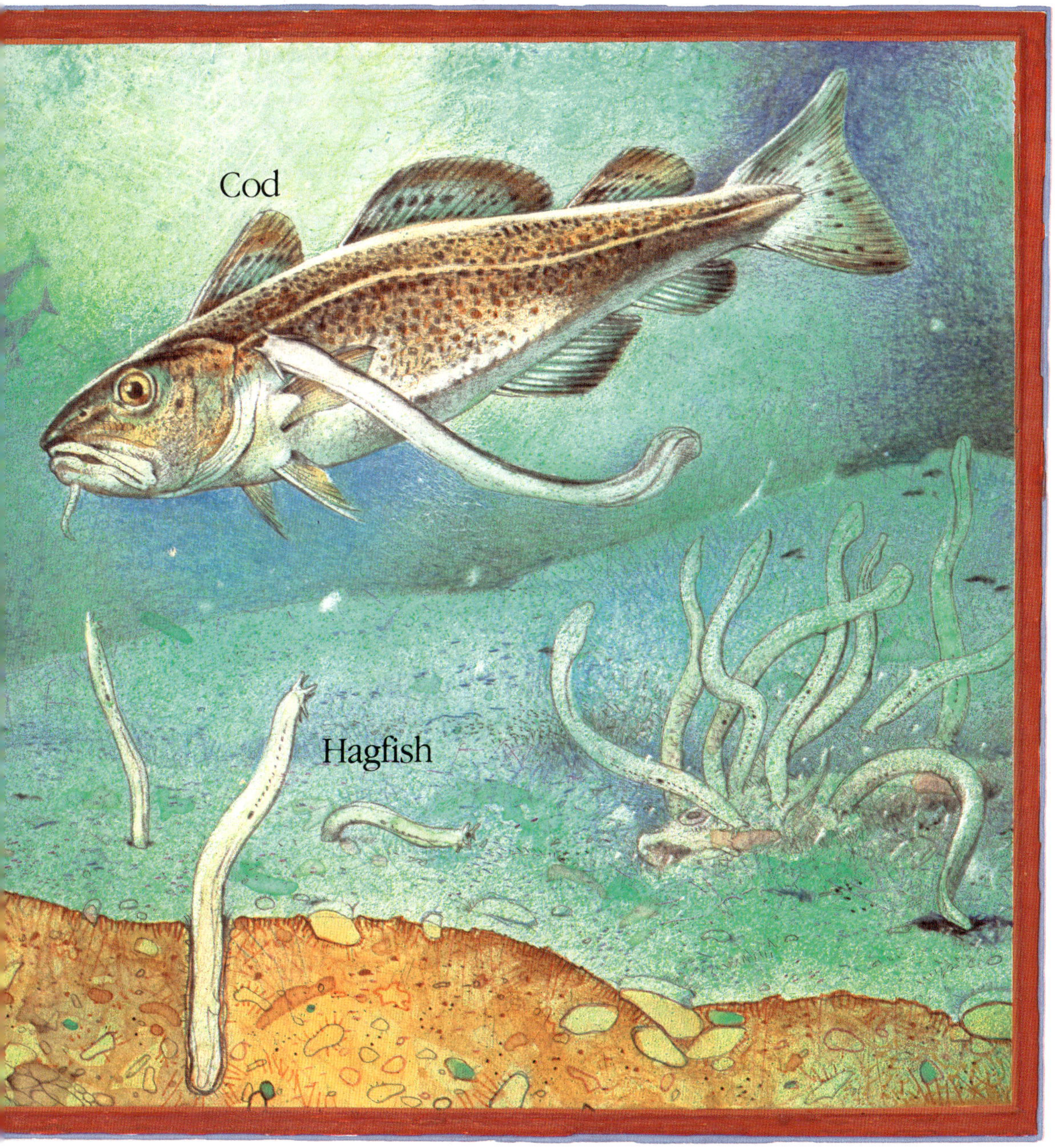

The cod is one of the victims of the glutinous myxines. Hagfish attack live creatures, especially if they are wounded or caught up in nets, but they also feed on the remains of dead animals.

THE MARINE DEPTHS

A veritable multitude of animal life thrives on the sea bed, patiently waiting for a little food to drop down from on high. The deeper the seas, the greater is this multitude at the various depths. And scraps from their meals, their excrement, and their bodies when they die all end up in a continuous shower of food sinking to the bottom. None of it goes wasted: the hagfish specialize in the larger prey, and other fish consume even the most minute particles of food using the old method of suction and filtration.

piece of the unfortunate victim. The hagfish swallows the food and begins again. In just a few minutes, these myxines are capable of hollowing out a true and proper tunnel inside their victims, and they disappear inside, continuing their devastating task.

Hagfish use the same unusual technique to escape from man: if you get hold of one in your hand, it will instantly go into the same knot that will slide from tail to head and enable the creature to slip out of your hand. The original reason for adopting the technique was not, however, to avoid being caught by man. It was because there was no other way of overcoming the impossibility of going hunting, we might say, with a round, unmovable mouth.

REGRESSION

In all animals, adaptation to parasitism causes a regression of their anatomical organization. Evolution in these forms concentrates only on the sense organs and the apparatus that enables them to be aggressive (as in the sucker mouth of the myxines) or to recognize their prey. All other organs, useless for this task, regress and can eventually disappear. If evolutionary selection does not continuously favor a given anatomical structure, that is, a function and its genetic information, the same structure tends to regress and then disappear.

4. EXISTING AGNATHA: THE PETROMYZON CYCLOSTOMES

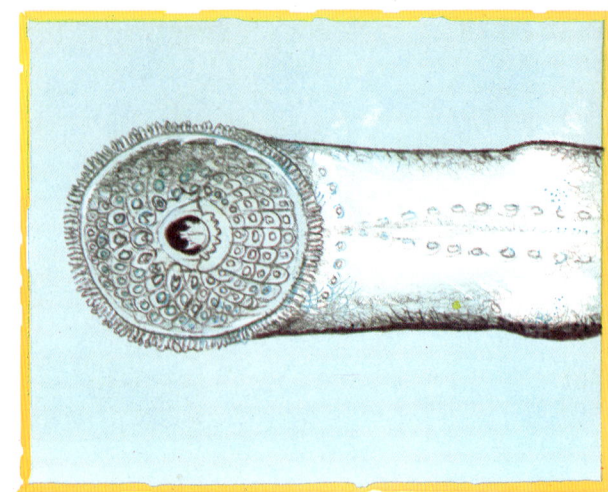

The lamprey's sucker-like, tooth-lined mouth with the tongue in the center.

THE LAMPREYS

Lampreys belong to the second order of *Agnatha* that has survived, the petromyzons. With these creatures, we see the emergence of rather refined behavior patterns and anatomical features. The body is eel-shaped, up to four feet long, and perhaps even longer. In the larval stage, the animal behaves in the old suction/filter way, but the adult is terribly aggressive. Its favorite prey are fish and cetaceans, members of the whale family. It attacks them ferociously and kills them swiftly.

ONE SOLUTION TO THE PROBLEM OF THE UNMOVABLE MOUTH

Lampreys too have to operate with a round, unmovable mouth, but they find a simple, efficient solution. The walls of the suction-type mouth are armed with extremely sharp, horn-like teeth and the back of the mouth is occupied by a muscular tongue that acts like a piston and is also covered with teeth. A fairly good swimmer, the lamprey darts rapidly at a passing prey and sticks its mouth to the body. The victim thrashes around trying to shake the attacker off, but the mouth cannot be detached; on the contrary, with all the movement, the lamprey's teeth begin to rasp the animal's body while the tongue pumps back and forth, sucking out the organic liquid as it forms. As if this were not enough, the lamprey's mouth contains glands producing a secretion that not only prevents the victim's blood from coagulating but actually helps the blood flow out of the prey's body. The poor creature soon bleeds to death while the lamprey continues his meal undisturbed.

THE GREED OF THE LAMPREYS

Lampreys are incredibly greedy. The story of what has happened in the Great Lakes of the United States in the past 50 or 60 years has rightly become famous. At the beginning of the century, the Lakes yielded some 3,900 tons of quality fish a year, trout and salmon mostly. But for reasons that are still not clearly understood, vast numbers of lampreys began to appear and, in just a few score years, the catch decreased to about 10 or 12 tons, all the rest being devoured by the ravenous cyclostomes.

One explanation lies in the bizarre sex habits of the lampreys.

HOW LAMPREYS REPRODUCE

Lampreys are born close to the sources of rivers, in fresh water, shallow and cold. After their metamorphosis, they descend from the river to the sea, where they spend years living the life of a parasite, off fish and cetaceans. When they are sexually mature, they stop eating and set out on the long journey that will take them back to where they came from. During their migration, they manage to survive formidable rapids and currents, using their sucker mouths to get a grip on rocks and flip from one to another. When, at last, they reach waters suitable for reproduction, they couple and lay their eggs. Now exhausted, they simply die.

The havoc wrought by lampreys in the Great Lakes could well derive from a change in their habits. Trapped by locks and dams, or perhaps for some other reason, they have accepted the Lakes as substitute for the sea and use the effluents as places for reproduction. Left safe to grow and mature in the Lakes, the lampreys reproduced undisturbed, one other reason being the absence of the natural enemies they would have encountered in the sea. Meanwhile, they quite comfortably massacred all the other lake dwellers.

One final word. Lampreys are delicious to eat, tender and delicate, food fit for a king, they say. But it is difficult to catch them, and you will never find them in your local fish shop.

PARASITISM

Parasitism is when one animal feeds and lives at the expense of another that receives nothing but harm or even death from the relationship.

Lampreys are good examples of parasites. Attaching their sucker mouths to their victims' skin, they feed off the underlying tissues. If the victim is big and strong enough, it need not always die.

In the outer frame: sea lampreys live in coastal waters, living off mackerel and other fish (1). In the mating season, they abandon the sea and make their way up the rivers (2). Lampreys even manage to suck their way up waterfalls by attaching their mouths to stones (3), or even to their fellow travellers, salmon. When they arrive at the right spot, they mate and the eggs are laid (4). The young lampreys stay in the river (5) until they are fully developed, at which point they set off for the sea (6).
In the center square: river lampreys spend their whole lives in fresh waters, though they do go on migrations. Here, one is attacking a rainbow trout.

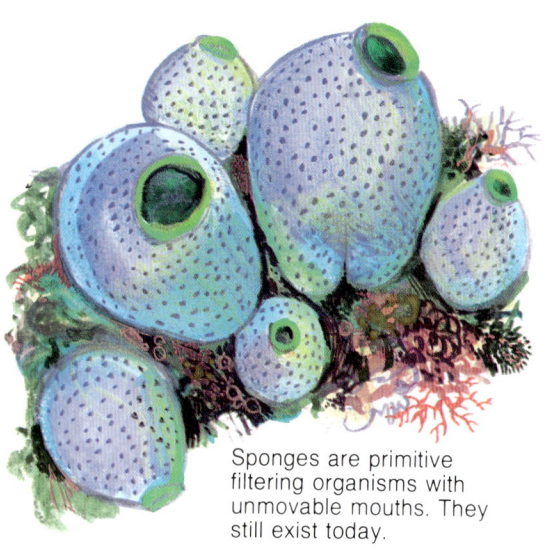

Sponges are primitive filtering organisms with unmovable mouths. They still exist today.

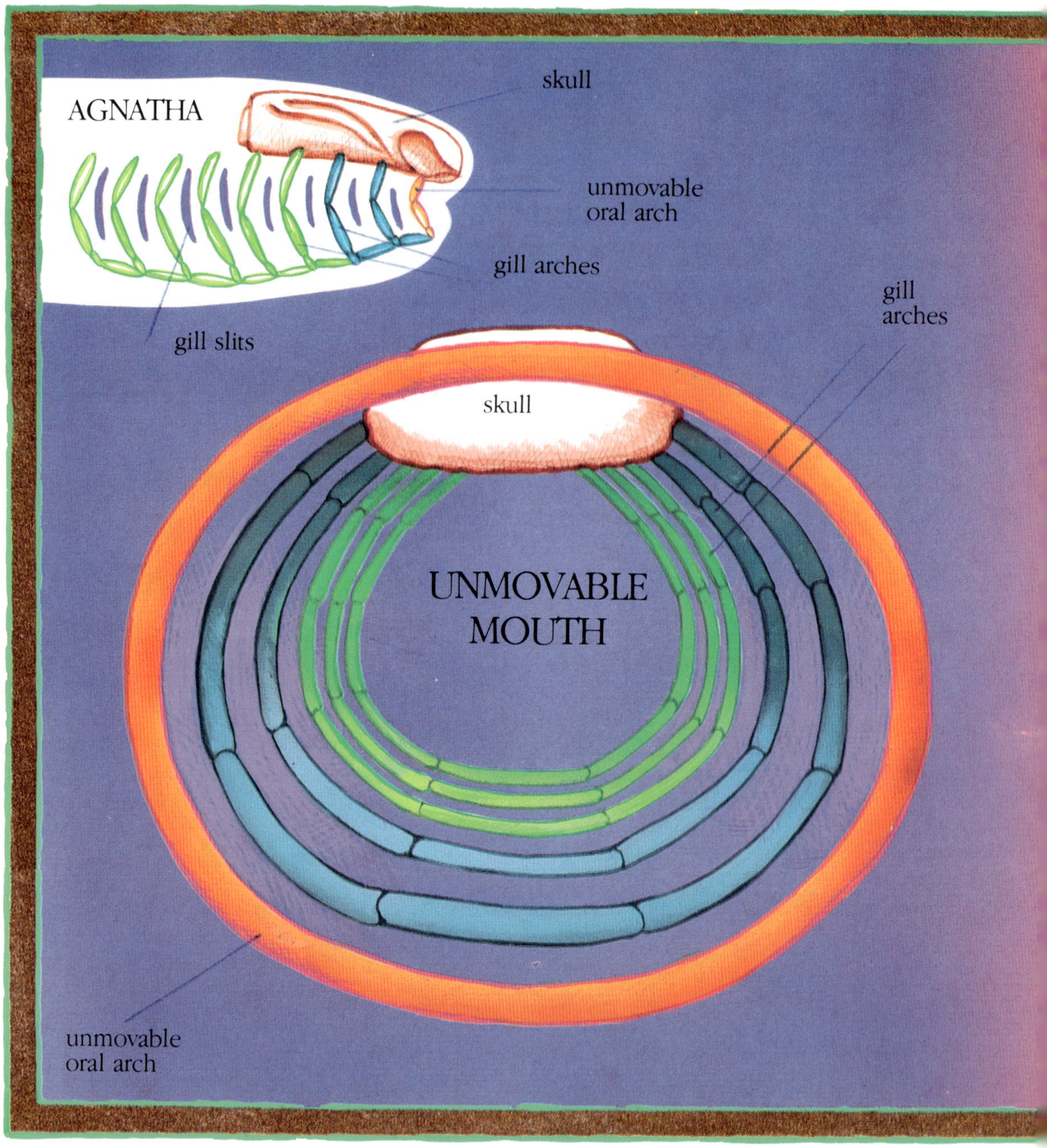

5. HOW A MOUTH FOR PREYING WAS INVENTED

About a billion years ago, evolution discovered that an organism consisting of a multitude of cells, single or in groups, each of which had special tasks to perform, was a far more efficient and functional organization than the sum of the capacities of all the cells put together.

This discovery soon led to the diffusion of a new form of life in all the Earth's waters: the multicellular organism. One of its most important and special tasks was catching prey to feed the whole organism.

FEEDING METHODS

The simplest way of feeding was to filter water and strain off the organic cells suspended in it. Still living on today are some very ancient invertebrates, the sponges, extremely simple sea creatures, that suck water in through their tiny pores and force it out through one large aperture after filtering it to retain oxygen and food, in the form of single-cell weeds, organic waste and bacteria.

But as evolution went on, the invertebrates very soon discovered that they needed a mouth and other special apparatus to enjoy different forms of food. The mouth did not take long to develop in this phylum: and when it came, it could grasp, suck, chew and crush.

Fish lagged way behind in this respect. Not only their ancient forerunners but also the whole of the first group of vertebrates were creatures that filtered water and mud for food. As we have seen, even though it did rule the Earth's waters for a very long period, and even though it developed very sophisticated forms, the ostracoderm never had more than an unmovable mouth that could do no more than filter.

A MOVABLE MOUTH FOR VERTEBRATES TOO

The movable mouth appeared quite suddenly in the mid-Silurian Period, more than 400 million years ago. It was a mouth with a mandibular arch, a jaw, that could open and close, and it had teeth. The mouth changed eating methods drastically. Animals could now go off and look for their prey, they could hunt and secure for themselves large quantities of food.

The discovery, if we can call it that, gave instant rewards. The new forms of fish soon began to take over the Earth's waters, to the detriment of the harmless ostracoderms, of course.

The breathing mechanism of a fish: water enters the mouth and, when the mouth is closed, flows into the gill chambers after the gill coverings have been raised.

GILLS AND BREATHING

Fish obtain the oxygen they need for breathing with their gills. The abundant mass of vessels in the gills not only helps to collect oxygen from water but also to give up carbon dioxide, just as in any other kind of respiration. The water always circulates in one direction — from the mouth to the gills — and this facilitates the exchange of the two gases. The fish uses the gill chamber to open and close its mouth in such a way that water enters it. By successively opening and closing the mouth and opening the operculum, or gill covering, the water is forced first into the pharyngeal intestine, then over the gills and finally to the outside.

Fish can extract a very large amount of oxygen from water, something like 92 percent at times (in the trout, for instance). However, this exceptional performance also makes the fish very sensitive to pollutants in the water, for they absorb pollutants just as easily. Fish are, in fact, used as biological indicators of water pollution.

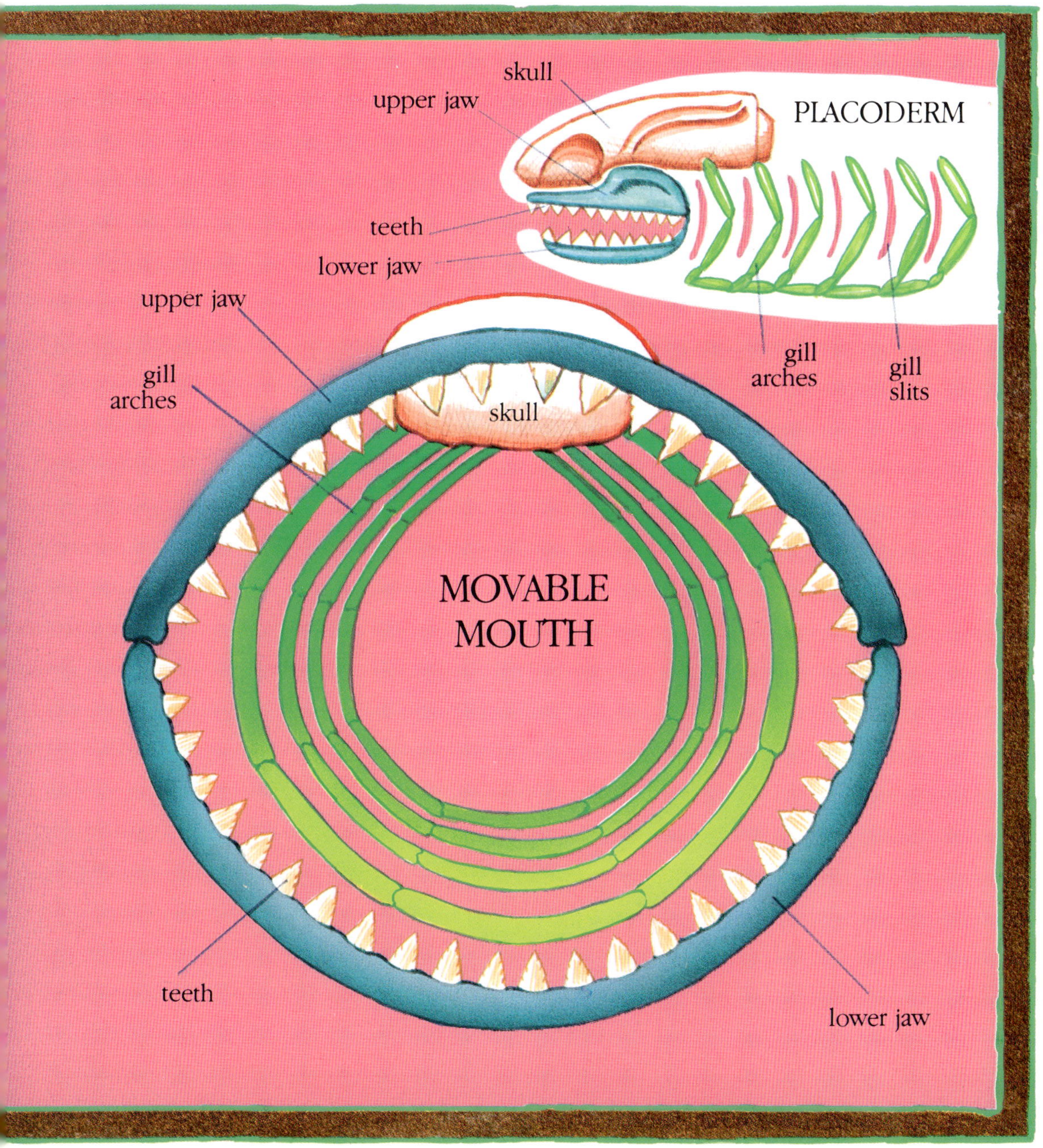

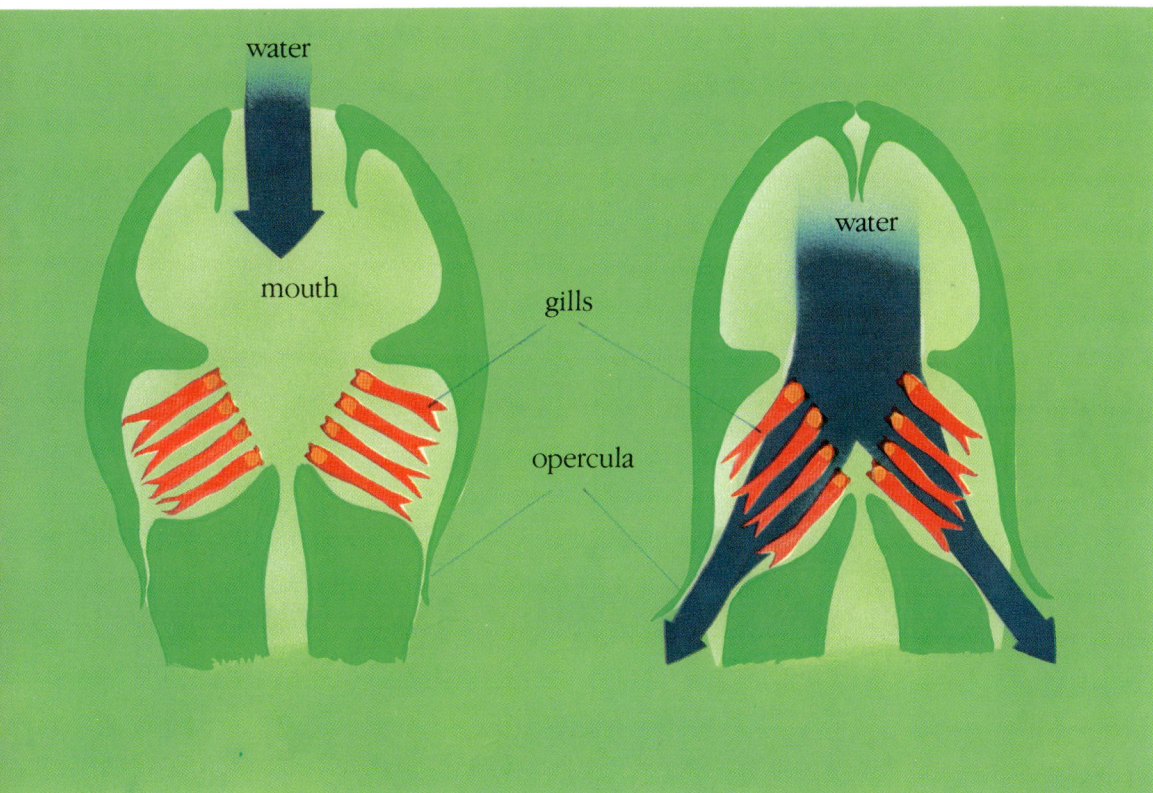

Reconstruction of a fossil of the *Dinichthys*, a gigantic placoderm that lived 350 million years ago. We see the powerful movable structure of the mouth, with upper and lower jaws.

THE MOVABLE MOUTH, A FUNDAMENTAL STAGE IN THE EVOLUTION OF THE VERTEBRATES

The newcomers were the placoderms (plated skins), and we shall have more to say about them in the next chapter. They did not differ very much from their ancestors, but they did have a fearsome mouth armed with exceptionally sharp teeth. The placoderms soon spread throughout all the Earth's waters, and no other creatures could challenge them. They also grew in length and weight; for the first time, forms of life in the waters became gigantic. Although they did acquire their movable, preying mouth quite late, the vertebrates grew from small organisms like the ostracoderm, a few inches long and a few ounces in weight, to creatures many feet long weighing hundreds of pounds.

With the placoderms, life took on a new, spectacular dimension.

WHERE THE MOVABLE MOUTH CAME FROM

There is an enormous difference between the round, unmovable mouth of the ostracoderms and the movable mouth armed with upper and lower jaws of the placoderms. Unfortunately, we know of no intermediate forms between the two that can help us to understand where the new one came from. There is only a hypothesis, which most accept because it is highly probable.

The unmovable mouth of a filtering organism like the ostracoderm was reinforced with a ring of cartilage. Behind this, there was a series of articulated, movable arches that supported the gills. As they rose and dropped, they allowed the creature to breathe.

The movable mouth may well derive from the second or third of the arches, for the mouth does consist of two half-arches linked together. They are the upper jaw, the maxilla, and the lower jaw, the mandible. The third and other arches retained the function of supporting the gills.

Animals endowed with this kind of movable mouth are called *gnathostomata*, from the Greek word that means "mouth with jaws."

11

Front view of a reconstructed *Rhinosteus*, showing the typical conformation of the placoderm's head, well protected by the bony outer armor-plating.

Gemuendina

6. THE FIRST VERTEBRATES WITH REAL MOUTHS: THE PLACODERMS

As we have seen, the inevitable decline and fall of the ostracoderms began with the appearance of new fish. These fish, like the cyclostomes, for example, were quite capable of attacking their prey, tearing it to pieces and swallowing it down naturally, that is, without the need for any special extras. The now-extinct newcomers, the placoderms, armed with upper and lower jaws, were the forerunners of all such other vertebrates, who are grouped under the term gnathostome, jaw-mouth, to give the right emphasis to the new acquisition. The question of the origin of these vertebrates is also fascinating. Just where did the placoderms come from? We have no intermediate forms to help us, so we can only guess.

PRIMITIVE PLACODERMS

At the beginning, the placoderms lived in fresh waters. Only later, when they were masters of all other forms of life, did they migrate and establish permanent colonies in the seas, too.

The first placoderms were decidedly similar to the older ostracoderms: the head and part of the front of the body were covered by a sturdy, bony armor-plating. This observation leads us to assume that the placoderms derived from very old, little-evolved forms of ostracoderm.

HOW EVOLUTION WORKS

Here we see an aspect of evolution that is to be repeated time and again. A new form, a new class, does not derive from the more advanced preceding forms; on the contrary, it originates from the more primitive organisms, less differentiated and less evolved.

The first placoderms did not resemble in any way at all the latest ostracoderms, whom they were to fight and wipe off the face of the Earth; they were definitely more similar to the earlier ostracoderms. So here, for the first time, we have an example of a fixed rule, if not a law, of evolution. It is the less specialized, less evolved forms that are flexible enough to follow the path of evolution.

THE MOUTH OF THE PLACODERM

So the placoderms first made their appearance in the fresh waters of the Silurian Period, some 400 million years ago.

They had a fairly easy time of it at the beginning. Their terrible mouths armed with sharp teeth gave them a tremendous advantage over the harmless ostracoderms. It was an unequal battle, if indeed there was a battle. The old lords of the sea gradually gave way to the newcomers. The placoderms took over the fresh waters first and then the seas, and that was the end of the ostracoderms. They all disappeared, and of the great group of *Agnatha*, only the cyclostomes have survived to this day.

THE RISE AND FALL OF THE PLACODERMS

The first fish with movable mouths began their period of maximum expansion during the Devonian Period. Some marine forms reached lengths of 20 feet, and not even the nicely evolved invertebrates had ever been as large as this.

And yet, the overwhelming might of the placoderms declined and fell. In the waters of the Devonian Period, many diverse, productive phyla of fish flourished and spread, and they were much better equipped for tackling the environment, more competitive and better able, therefore, to colonize different kinds of water environment.

As if under some biblical condemnation, the placoderms first conquered the ostracoderms they derived from and were then conquered by their own descendants, relentlessly crushed by the impetuous rise of more modern forms.

Rhamphodapsis

Dinichthys

Pterichthys

THE ACANTHODIANS

All we have of these strange fish, now extinct, are their fossilized imprints. They are a source of embarrassment for the specialists. Some maintain the acanthodians (spiny fish) were a slightly different form of placoderm, while others place them in a group of their own. What seems certain is that they were the originators of the great phyla of modern fish that rose in the Devonian Period, spread thoughout all the Earth's waters, and survived. In other words, they are the ancestors of all the fish that followed.

The waters of the Devonian, some 350 million years ago, were inhabited by placoderms. Thanks to their movable mouths equipped with strong jaws and teeth, these hunters took over dominion of the seas from the ostracoderms - with their unmovable mouths — and then disappeared in turn.

COMPETITION

The competition between ostracoderms and placoderms is a fine example of how evolution works. The same territory, the same ecological niche, or living space, is at a certain moment occupied by two great phyla, one very old, but primitive, the other more modern and efficient. The two forms cannot coexist; one has to chase the other away. As we shall see, competition is very common, but not the exclusive method of evolution. Other ways and means will allow forms to follow one another.

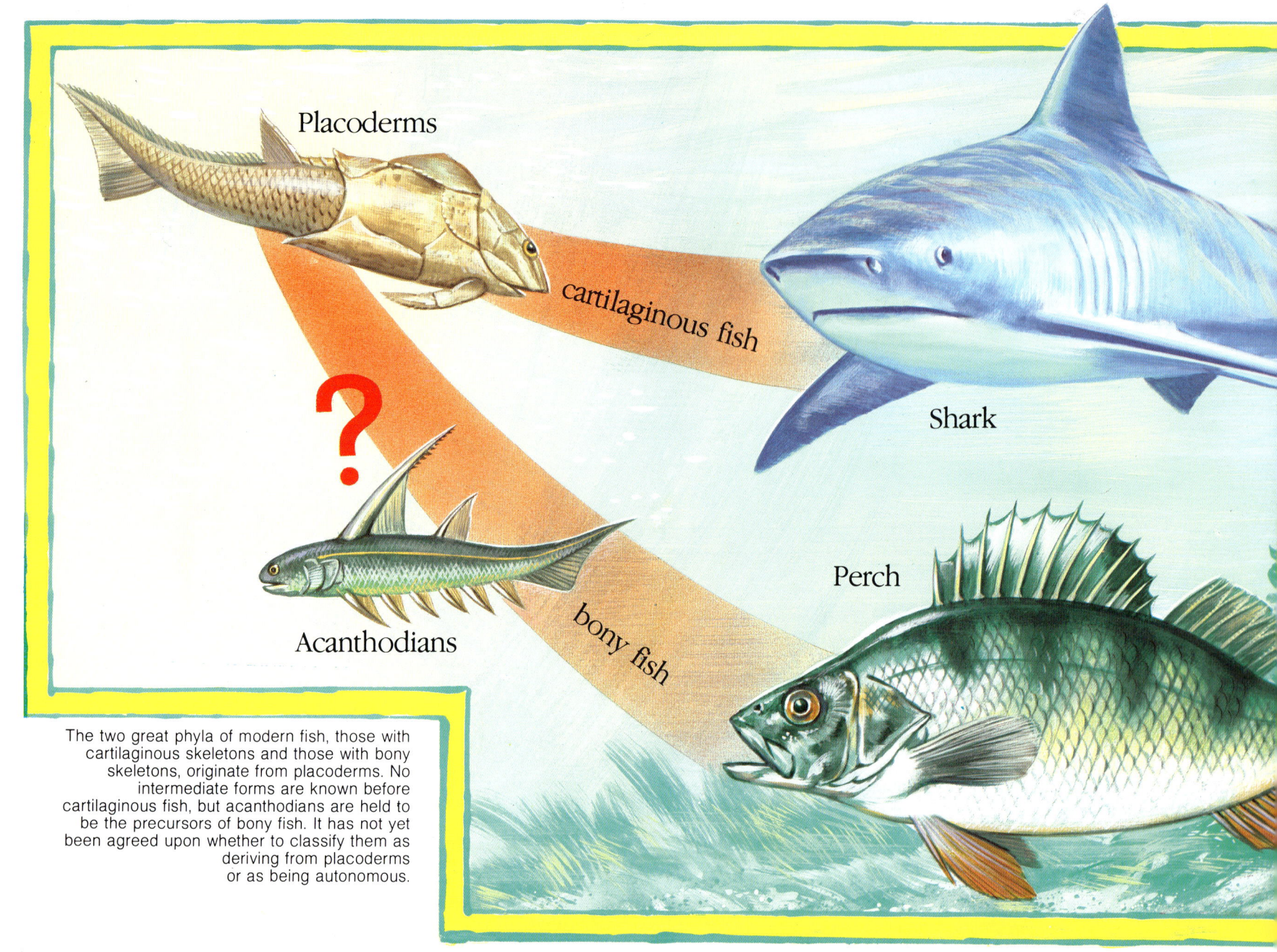

The two great phyla of modern fish, those with cartilaginous skeletons and those with bony skeletons, originate from placoderms. No intermediate forms are known before cartilaginous fish, but acanthodians are held to be the precursors of bony fish. It has not yet been agreed upon whether to classify them as deriving from placoderms or as being autonomous.

7. THE DEVONIAN PERIOD: THE GREAT AGE OF FISH

Every class of vertebrate has had its heyday, a period of great expansion in number of forms and variety of species. We shall see the splendor of the amphibians, reptiles and mammals in other volumes in this series.

THE AGE OF FISH

The Devonian Period is quite rightly known as a time of great fortune for fish. In fresh waters first and then in the seas, the placoderms, both minute and gigantic, were the unchallenged masters. But the most important event for the continuation of our story was the birth, if we may use that word, of two new, great phyla of fish. The placoderms were, in fact, to disappear forever around the end of the Carboniferous Period, some 300 million years ago, but from them, as we shall see, sprang two completely different lines of fish that were to follow their courses of evolution quite independently.

CARTILAGINOUS FISH AND BONY FISH

Fish with skeletons composed of cartilage inside their bodies derived from a form we still know little about. In turn, the shark family derived from them. Meanwhile, the history of bony fish, like trout and tuna, the most typical kinds of fish, began with the acanthodians. The whole event is quite curious, really.

The placoderm was a fish with heavy armor-plating protecting its head and some of the front part of its body, while its internal skeleton — the backbone and fins — was somewhat frail, made partly of cartilage and partly of bone. A more elegant, graceful fish derived from it; the armor-plating disappeared and the internal skeleton became stronger, but in two completely different ways.

In one phylum, or line, the skull and backbone were cartilaginous.

In the other, the skeleton was composed of bone tissue. One cannot say which of the two is the best. Both phyla developed huge forms and the different skeletons supported them perfectly.

Thus, cartilaginous fish, or *chondrichthyes*, and bony fish, or *osteichthyes*, derived in quite different ways from the placoderms of the Devonian Period and pursued their paths of evolution quite independently.

Living in water, as we shall see, means a large variety of problems have to be solved: osmotic insult,

cartilaginous skeleton

bony skeleton

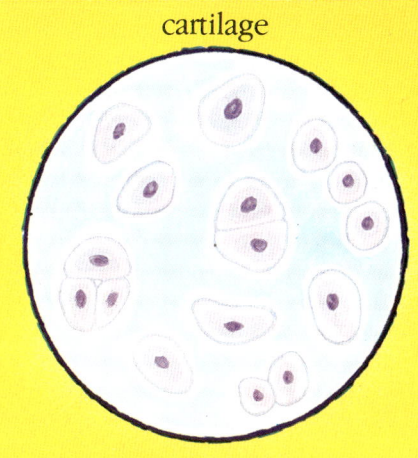

SUPPORTING TISSUE

A voluminous organism must of necessity have a supporting structure, a skeleton, around which it can wrap the muscles that enable it to move, and on which it can shape its body. In vertebrates, the task is entrusted to two different tissues: *cartilage* and *bone*. Both are composed of characteristic cells, fibers and a fundamental substance that enriches itself with salts and gives the tissue the stiffness it needs. The two types of skeleton can live together, but, as we shall see, evolution tends to favor bone, which is more plastic and dynamic than cartilage.

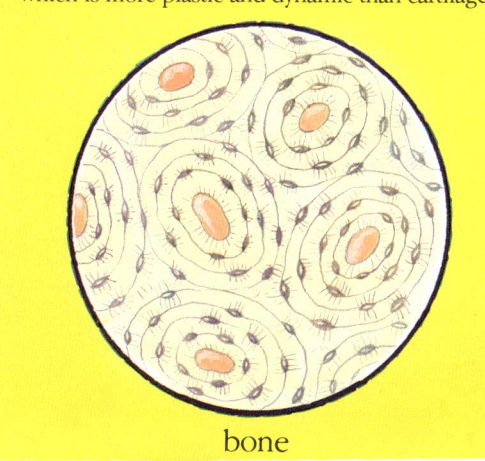

CONVERGENCE

One of the most characteristic phenomena of evolution is convergence of form. The same environment and the same requirements produce identical forms. Our two great groups of fish, cartilaginous and bony, were not the only creatures that had to adapt their bodies to cope with life in the water; the reptiles and mammals had to face the same problem. And the end result was the same: all had practically the same form, so much so that their overall appearance is not enough to distinguish them one from the other.

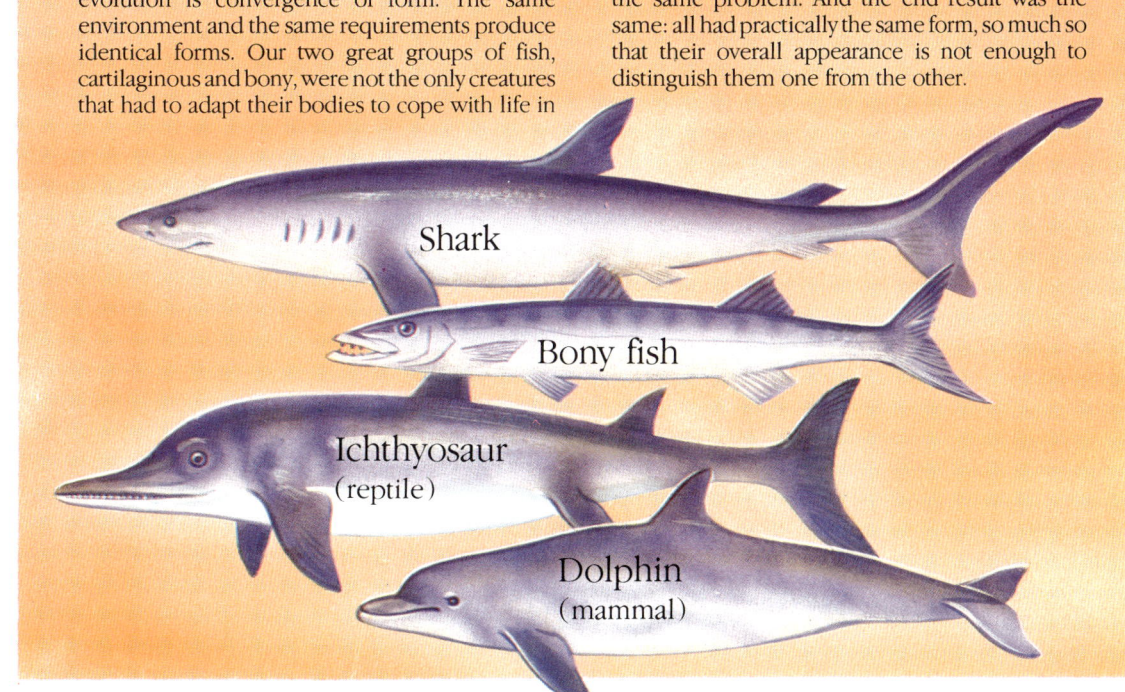

specific weight, the dynamics of swimming, reproduction. We shall see them all in turn. To show how independent of each other the two phyla were, and to show just how imaginitive evolution is, each group solved its problems in its own way, with the best solutions sometimes in one group and sometimes in the other.

The only feature they do have in common is their basic shape, modeled to meet hydrodynamic requirements: a streamlined body, powerful tail muscles for swimming and one or two pairs of stabilizing fins to correct rolling and pitching movements. But while their outward appearance is quite similar, practically everything we find inside the two groups is different.

To be able to live, fish need a constant balance between the concentration of salts inside the body and the concentration of salts in the outside environment (1). In salt water, the concentration of salts is higher than in the body of the fish and, on account of osmotic pressure, water flows out of the cell, which dries up (2). In fresh water, the concentration of salts is lower than in the cells and, on account of osmotic pressure, water flows into the cell, which eventually bursts (3). If it were not protected by a skin that was suitable for its environment, a fish in salt water would lose most of its water and its tissues would literally dry up (4). The same fish in fresh water would swell to bursting point and actually burst (5).

8. HOW CARTILAGINOUS AND BONY FISH SOLVED THE PROBLEM OF OSMOTIC PRESSURE

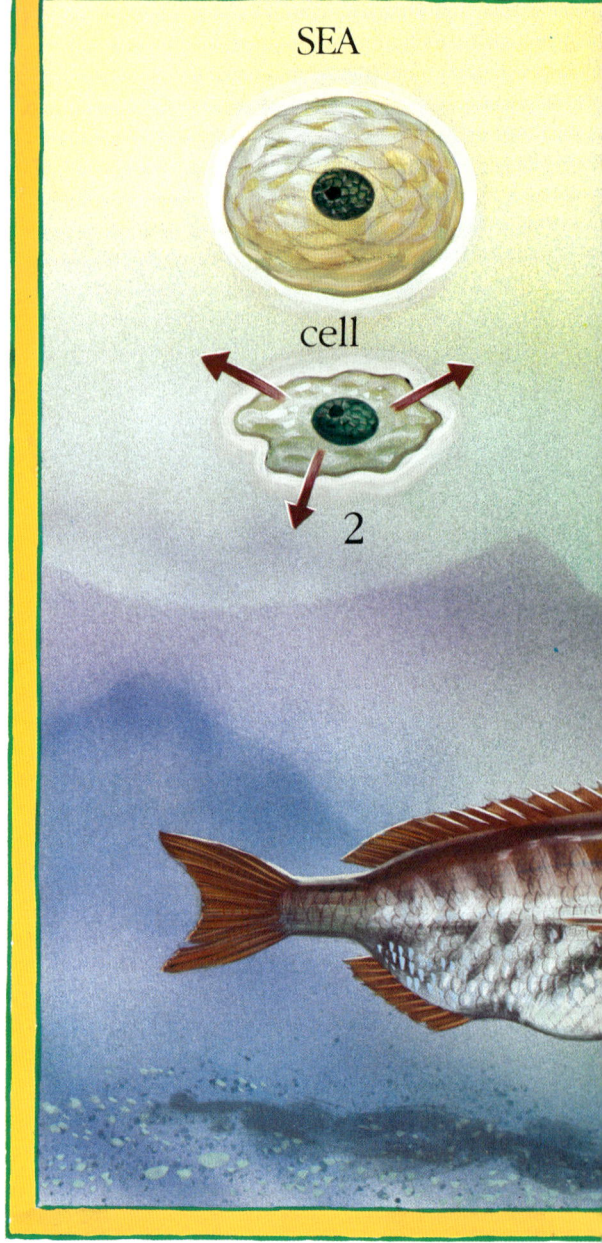

OSMOTIC PRESSURE

In a receptacle, let us prepare two solutions of a salt in different concentrations and separate them with a membrane that lets water through but not salt. In these conditions, the different concentrations create a pressure, called osmotic pressure, which causes water to flow from the less to the more concentrated part. The flow will cease when the two concentrations are equal and, in the end, the two solutions will have different levels.

THE PROBLEM

Osmotic insult is the most urgent problem facing an organism living in sea or river waters. Any cell, like any organism, contains a certain concentration of salts that may well be different from the concentration of salts on the outside.

A FISH IN THE SEA

A sea fish, for instance, contains a quantity of salts in its blood, muscles and tissue that is much lower than the quantity of salts in the water it lives in. Incidentally, if they were equal, the fish would be so salty that it would be uneatable. The difference between the two concentrations creates a heavy osmotic pressure, as it is called, that tends to balance out the two concentrations. That is, a flow of water is created that issues forth from the cells and tissues of the fish and tries to equal out the quantities of salts. As it comes out, the water increases the concentration of salts contained in the animal and dilutes the external salts at the same time. It may seem paradoxical, but without any means of defending itself, a fish in sea water would actually dry up! Water would flow out from all its cells in an attempt to make the fish salty, like its environment, and this would reduce it to no more than a practically empty container enclosing a skeleton.

A FISH IN FRESH WATER

Exactly the opposite happens with a fish in fresh water. The concentration of salts in its cells is higher than that of the water it lives in, so the flow of water goes in the opposite direction. From the outside, it penetrates cells and tissue in an attempt to dilute the salts in the body of the fish. In this case, the creature inflates like a ball, all its cells swell to bursting point and, indeed, it dies as all its cells explode together. This is the essential situation. The problem is to invent some way of preventing these incoming and outgoing flows of water. Let's see how fish do this.

THE SOLUTION FOR CARTILAGINOUS FISH

The chondrichthyes (cartilaginous fish) found a brilliantly efficient solution. All animals eliminate as waste matter a nitrogenous molecule. This may be ammonia, urea or uric acid. It is continuously eliminated with urine. Chondrichthyes, sharks and rayfish keep as much urea in their bodies as is necessary to balance out their concentration of salts with the salts in the sea. So these fish are in a state of osmotic equilibrium thanks to the high concentration of urea in their tissues. Shark fish meat does not taste salty when we eat it. Urea does not have that effect. At the most, only a slight odor, perceived by only the most refined sense of smell, gives away the origin of the fish.

This system for blocking the outflow of water that would otherwise dry the fish up is extremely efficient, but it cannot work the other way around; that

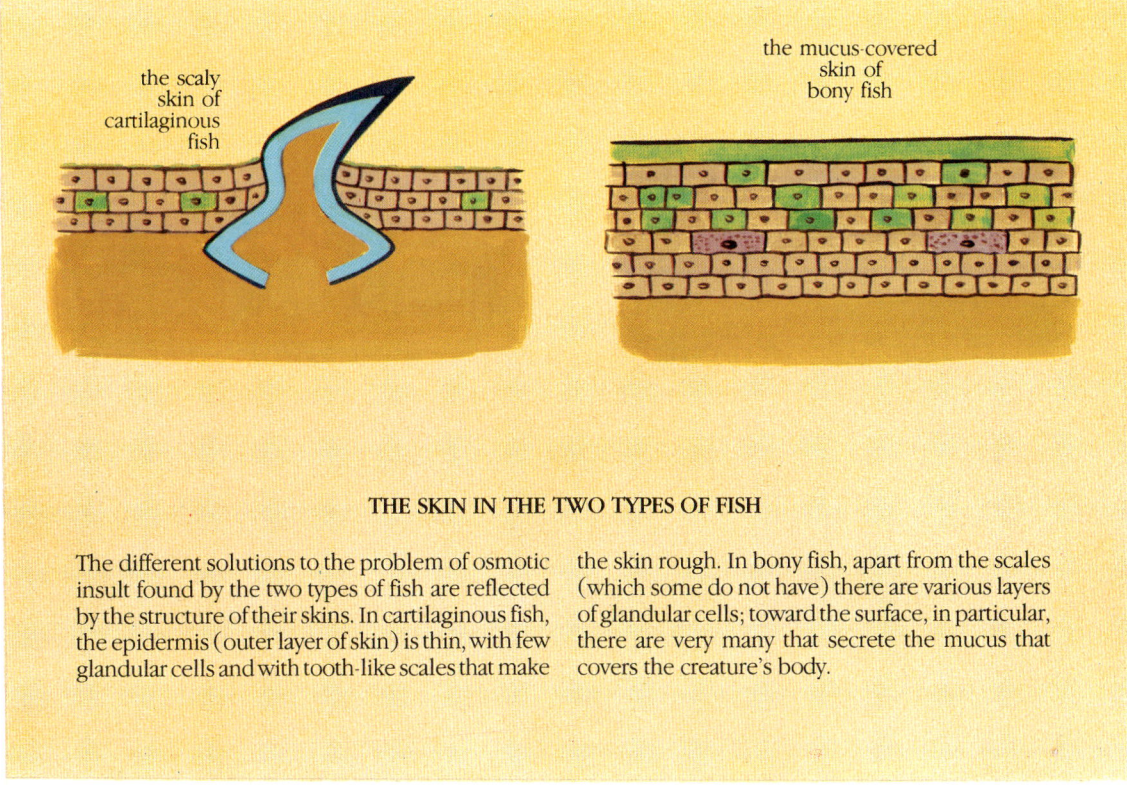

is, it cannot protect the creature in fresh water. This is, perhaps, one of the main reasons why most sharks and rayfish are saltwater fish.

THE SOLUTION FOR BONY FISH

In the phylum of the osteichthyes (bony fish), the problem was solved by covering all the surface area of the body that comes into contact with water — skin, mouth and intestine — with a waterproof substance, mucus, secreted by glandular cells. The film of mucus prevents water from entering and leaving the body of the fish, and the system works perfectly well both in fresh water and in sea water. The difference between the two kinds of fish can be told on touch. While cartilaginous fish generally have a roughish skin, bony fish are slippery, difficult to hold. They slip out of your hand because of the film of mucus that covers them.

THE SKIN IN THE TWO TYPES OF FISH

The different solutions to the problem of osmotic insult found by the two types of fish are reflected by the structure of their skins. In cartilaginous fish, the epidermis (outer layer of skin) is thin, with few glandular cells and with tooth-like scales that make the skin rough. In bony fish, apart from the scales (which some do not have) there are various layers of glandular cells; toward the surface, in particular, there are very many that secrete the mucus that covers the creature's body.

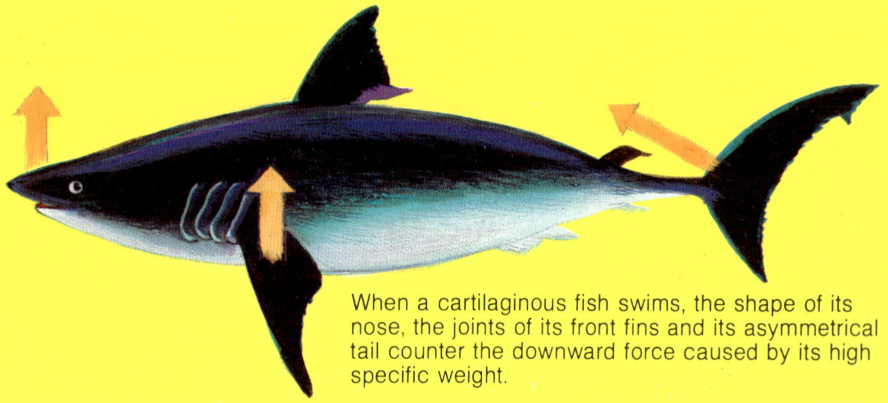

When a cartilaginous fish swims, the shape of its nose, the joints of its front fins and its asymmetrical tail counter the downward force caused by its high specific weight.

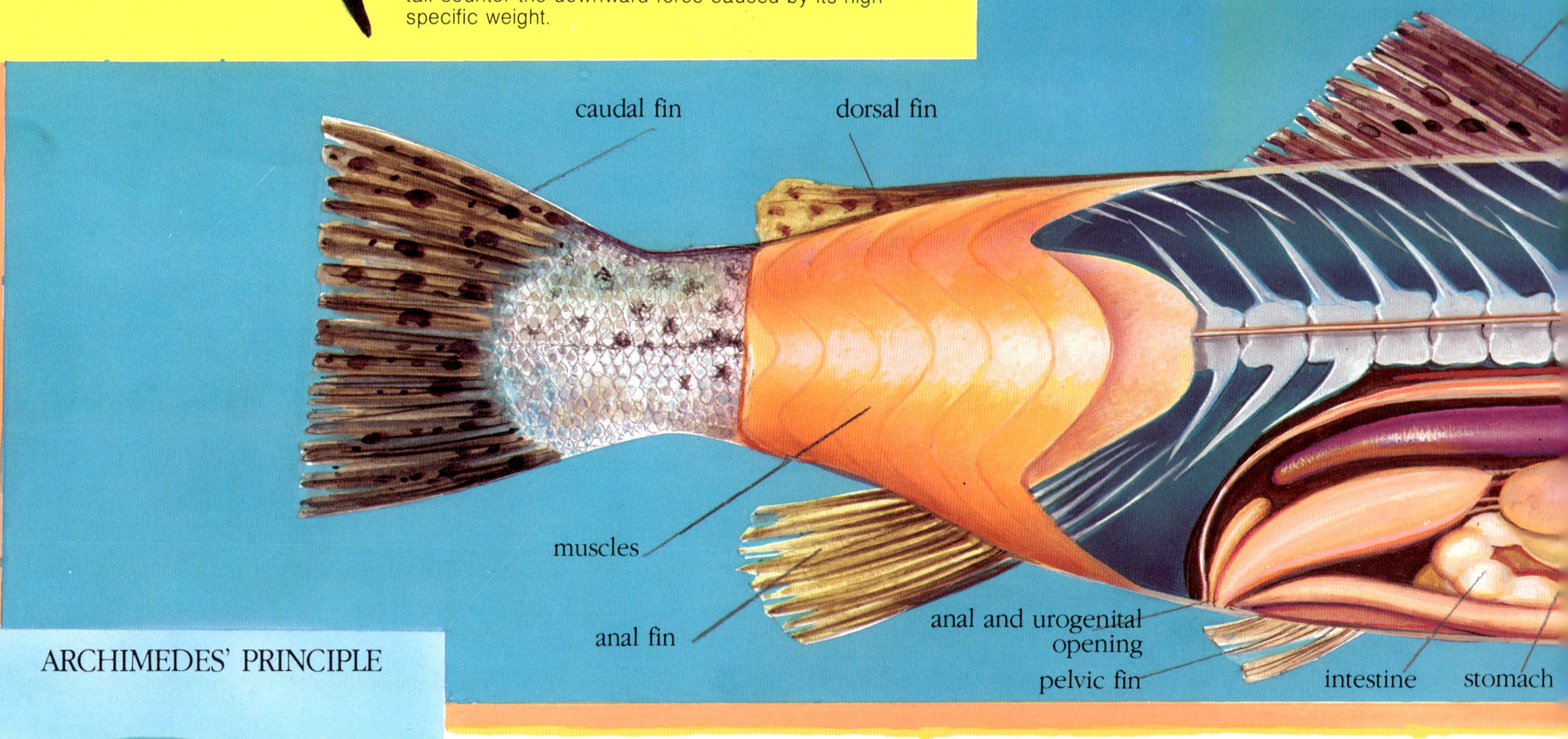

caudal fin — dorsal fin — dorsal fin

muscles — anal fin — anal and urogenital opening — pelvic fin — intestine — stomach

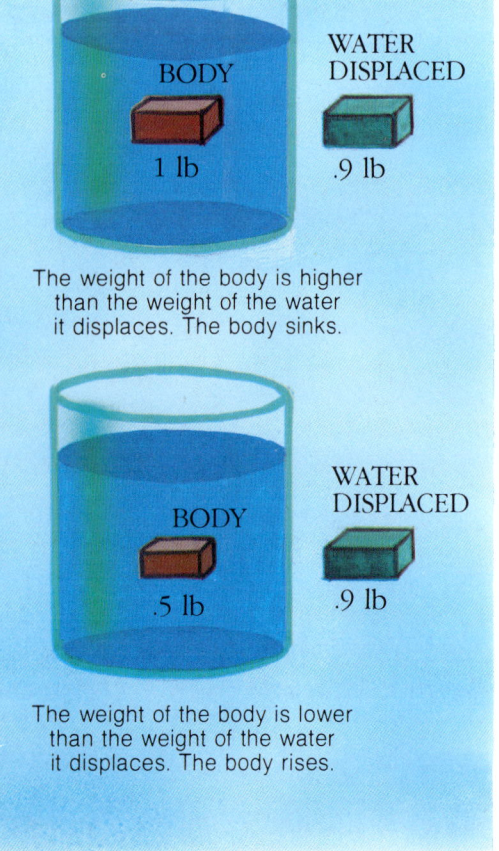

ARCHIMEDES' PRINCIPLE

BODY 1 lb — WATER DISPLACED .9 lb

The weight of the body is higher than the weight of the water it displaces. The body sinks.

BODY .5 lb — WATER DISPLACED .9 lb

The weight of the body is lower than the weight of the water it displaces. The body rises.

9. HOW THE TWO TYPES OF FISH SOLVED THE PROBLEM OF SPECIFIC WEIGHT

Living and swimming in shallow waters is not particularly problematic. But serious problems do occur in deep waters.

ARCHIMEDES' PRINCIPLE

To understand the difficulties, we should have a look at Archimedes' well-known principle. A body in water is driven upward by a force equal to the weight of water that body displaces. At the same time, however, gravity pulls the body downward at a force equal to the weight of the body.

One pound of lead displaces .9 pounds of water, and two forces act on it: one pulls it downward with a force of one pound and the other pushes it upward with a force of .9 pounds. So the lead inevitably sinks to the bottom. A pound of wood, on the other hand, displaces much more than a pound of water, so it floats.

Fish too have to deal with Archimedes' principle, the problem being further complicated by the fact that the volume of a fish's body can vary according to the pressure of the water.

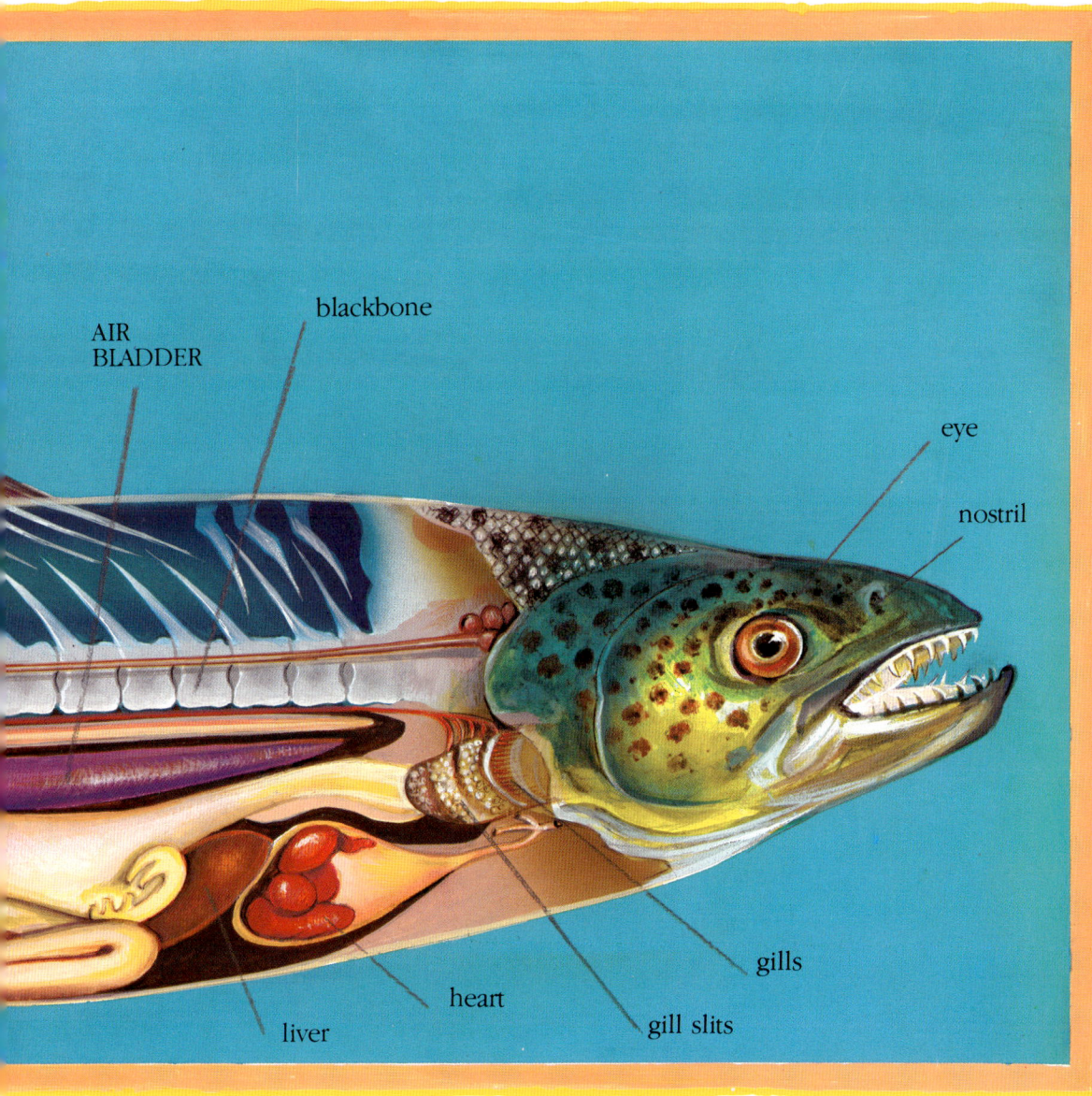

The anatomy of the bony fish contains the particular novelty of the air bladder. Full of gases, it helps to keep the specific weight of the fish equal to the specific weight of the water the fish displaces.

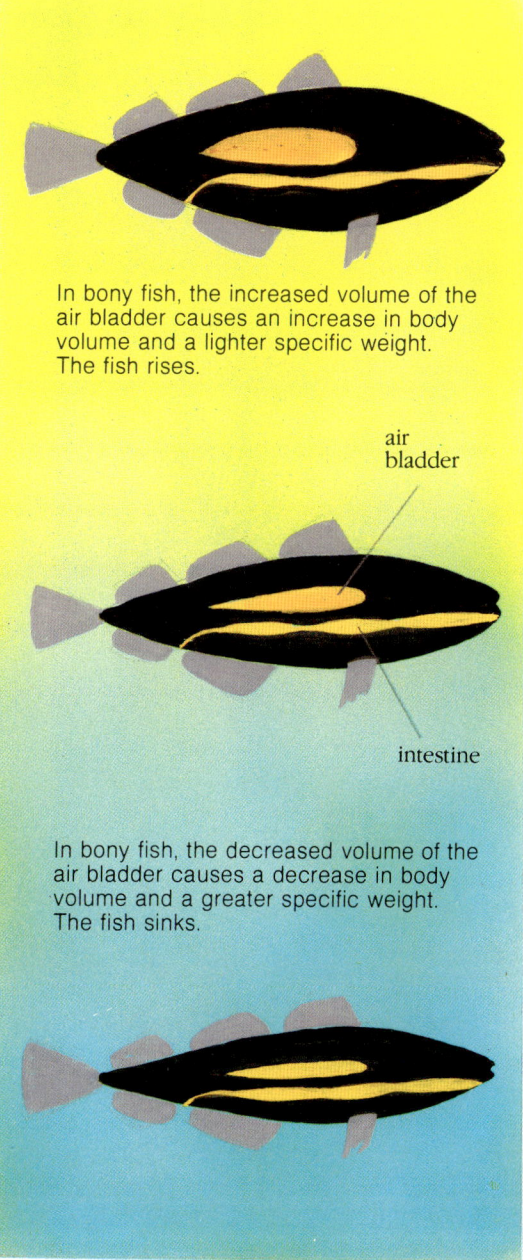

In bony fish, the increased volume of the air bladder causes an increase in body volume and a lighter specific weight. The fish rises.

In bony fish, the decreased volume of the air bladder causes a decrease in body volume and a greater specific weight. The fish sinks.

THE PROBLEM

Let us take an example. Specific weight is the ratio between weight and volume. The specific weight of a fish at a certain depth is equal to the weight of the water is displaces. So the creature is quite comfortable; it is forced neither upward nor downward. But if we shift the fish, toward the surface, let's say, the volume of its body increases and the fish is consequently forced upward because its specific weight is now lower. If, on the other hand, we shift the fish downward, the greater pressure of the water increases the specific weight of its body and the fish sinks to the bottom.

THE SOLUTION IN CARTILAGINOUS FISH

Rayfish do not have this problem; they usually live comfortably stretched out on the bottom. Sharks and the like, however, generally have high specific weights and have to overcome the force that keeps trying to drag them to the bottom. The solution lies in the shape of the nose, the layout of the front fins and the asymmetrical shape of the two flaps of the tail. As it swims, the shark generates not only forward thrust but also small upward thrusts that counter its high specific weight. So the problem is solved, but there is a great disadvantage in the solution. Sharks live in deep waters well off the coast and they can never have a break. They have to keep moving all the time. If they stopped for a rest, they would instantly sink to the bottom.

THE SOLUTION IN BONY FISH

Bony fish or, rather, the more evolved ones, like the teleosts, have a bladder containing a mixture of gases called the air bladder. It is used to keep the specific weight of the fish in a permanent state of balance with the specific weight of the water its body displaces. When the fish moves upward or downward, it simply has to reduce or increase the gases in the bladder to reduce or increase the volume of its body, thus keeping the two weights equal.

This is the ideal solution because the fish can stop for a rest, and it can move upward or downward.

10. FISH AT THE END OF THE DEVONIAN PERIOD

In the preceding chapters, we tried to stress the great differences between the two lines of evolution of the fish — the chondrichthyes and the osteichthyes — that appeared on the Earth's surface during the Devonian Period, some 400 million years ago.

THE END OF THE PLACODERMS

Like the placoderms, the two new phyla of fish probably originated in the fresh waters of rivers and estuaries and only later made their way toward the seas. Cartilaginous fish above all settled stably in the seas, where they made life uncomfortable for the old placoderms. These ancient armored fish therefore had to tackle competition on two fronts: in the seas, with the cartilaginous fish and, in fresh waters, with the bony fish. It is somewhat difficult to imagine what a fight between a placoderm and a primitive shark must have been like; it was probably a question of better efficiency. Both forms were predatory fish with mouths and teeth suitable for the purpose, so we can assume that victory went to the one that preyed more, the one that managed to deprive the other of food. There was probably no direct encounter, because the placoderms had reached massive sizes during the Devonian Period, while the cartilaginous fish were much smaller. But their greater swimming agility, more rapid reproduction, and better adapted mouths swung the advantage over to the cartilaginous fish, and that was the end for the placoderms. Slowly, but inevitably, one after the other, they began to disappear.

The dominion of the first great group of real fish came to a finish toward the end of the Carboniferous Period, when the chondrichthyes and osteichthyes had already taken command of their favorite territories.

No member of the great group of placoderms that populated our Earth for over 100 million years has lived on. The placoderms were the only great group of vertebrates that left no living trace of itself; none survived.

In the sea, in spite of their size, those last placoderms had to give in to the first ferocious cartilaginous fish.

Starting with the acanthodians, perhaps the forerunners of all bony fish, the bony-fish phylum split into two: the *Actinopterygii* inhabited the deeper, cooler waters, while the *Sarcopterygii* lived in warm, shallow waters, with little oxygen.

ADAPTATION

Here, we have a clear example of adaptation. The different environments — one with plenty of water and oxygen, the other with warm marsh water with little of the precious gas — give way to different anatomies and functional capacities.

As evolution went on over the millions of years, fish became ever better adapted to their environment, and their general organization, very similar at the beginning, changed profoundly.

TWO LINES OF EVOLUTION IN THE BONY FISH

The waters of the Devonian Period also witnessed another very important event. While the line of cartilaginous fish went on developing quite independently, there was a fork in the bony fish line. To understand why, we have to recall that, in those days, the mainland of the Earth, if we can call it that, was mostly marshland, shallow lakes and ponds. The temperature was also extremely high, so there cannot have been much oxygen available in the water. At that time, bony fish split into two great phyla: one stayed on to colonize cool waters, deep and with plenty of oxygen, while the other began to adapt to life in the marshes. In marshy environments, respiration through gills is no longer sufficient because there is not enough oxygen in the water, so creatures had to learn how to use the oxygen in the air as well. And then, rather than swim, they had to learn how to crawl, creep and slither.

In brief: one group of osteichthyes, the *Actinopterygii* (ray-finned) went on with evolution as typical fish, while another, the *Sarcopterygii* (flesh-finned) adapted to life on the boundary between land and water. And this was extremely important for the rest of our story.

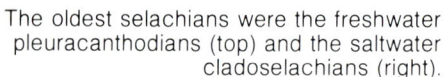

The oldest selachians were the freshwater pleuracanthodians (top) and the saltwater cladoselachians (right).

Turtle

11. CARTILAGINOUS FISH, OR THE CHONDRICHTHYES

THE SELACHIANS

The first fish with cartilaginous skeletons appeared in the Devonian Period. They were the chondrichthyes, and our present-day sharks (selachians) and rayfish derived from them. The precursors of the class were not very large creatures, never much more than four feet long. But from the very beginning, the shape of their bodies, their powerful tails and broad fins made them excellent swimmers. And their strong, well-armed mouths made them exceptionally mighty, dangerous attackers. They struck it rich right away; they soon spread through the salt seas and fresh waters, where they took over from the heavier, stubbier placoderms, and where they came into competition with the other phylum of existing fish, the osteichthyes, that is, fish with skeletons made of bone tissue. The selachians did not expand in the same way in the two environments: in the fresh waters, they had to yield ground to other fish, while in salt waters, they made their mark immediately, with different and sometimes gigantic forms and dimensions. From the moment they arrived, over 350 million years ago, until the end of the Triassic, less than 200 million years ago, the selachians continued to expand in numbers, families and species. It seemed that their reign would never end, but it was right then, after the Triassic, in the middle of the Jurassic, that a new threat unexpectedly stopped their expansion or, rather, slowed it down.

CONFLICT WITH THE REPTILES

The selachians had had an easy victory over the placoderms and had established a reasonable agreement with the bony fish, but they could not get on with the newcomers, the reptiles, who were beginning to leave the mainland and come and live in the seas too. The Secondary Era, rightly called the Reptilian Era, was characterized by the vast expansion of these vertebrates, who developed various new phyla in the seas, all of them adapted to the new environment.

MARINE REPTILES

Alongside the old turtles that still live on today, the seas were full of plesiosaurs, gigantic preying reptiles with squat bodies and long, slender, rather elegant necks. There were also the mesosaurs, one of which was the *Tylosaurus*, a reptile with the shape and size of our crocodiles (but the two are not related); with their powerful mouths armed with very sharp teeth, they must have been formidably aggressive competitors. Then there were the ichthyosaurs, the other phylum of marine reptiles: they reached the same size and shape as our cetaceans (dolphins and whales) and reigned over the seas for a long time in the Jurassic Period. All these forms of reptile, well-adapted to the marine environment, caused a drastic reduction in the number of selachians and slowed down their expansion. Indeed, both direct

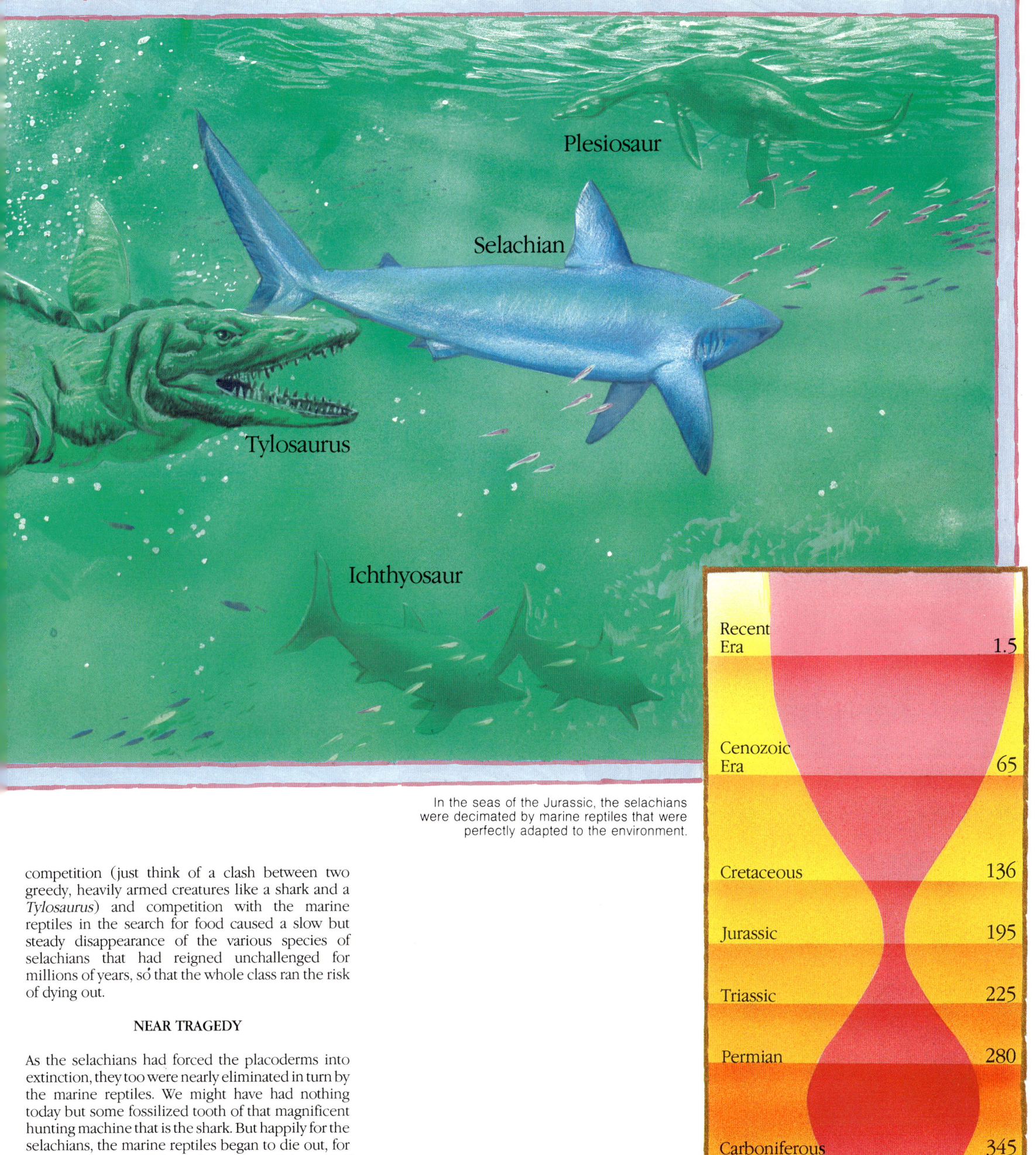

In the seas of the Jurassic, the selachians were decimated by marine reptiles that were perfectly adapted to the environment.

competition (just think of a clash between two greedy, heavily armed creatures like a shark and a *Tylosaurus*) and competition with the marine reptiles in the search for food caused a slow but steady disappearance of the various species of selachians that had reigned unchallenged for millions of years, so that the whole class ran the risk of dying out.

NEAR TRAGEDY

As the selachians had forced the placoderms into extinction, they too were nearly eliminated in turn by the marine reptiles. We might have had nothing today but some fossilized tooth of that magnificent hunting machine that is the shark. But happily for the selachians, the marine reptiles began to die out, for some reason we still do not know, toward the end of the Secondary Era, and selachians resumed the expansion that made them masters of the seas.

The expansion of the selachians over the eras and periods, in millions of years.

The effect of the dorsal position (toward the top of the body) of the eyes and nostrils of bony fish and the ancestors of the sharks was that the fields of observation of the eyes and nostrils overlapped a good deal and did not include the area below the body.

12. MIGRATION OF THE MOUTH AND SENSE OF SMELL IN THE SHARK

The story of the evolution of the selachians is a static one.

When they first appeared, in the Devonian Period, they were already full-fledged. With the passing of millions of years, from 350 million years ago until today, they have changed very little.

Only one thing really did change a lot: the mouth, from its position under the end of the snout, migrated ventrally, that is, downward, toward the belly, and took up a decidedly awkward position as far as eating is concerned. But if this happened at all, it must have given the selachians some advantage. Before we go any futher, let us see something of the creature's sense organs.

THE DOMINANT SENSE ORGANS

In present-day fish as in ancient fish, three sense organs dominate over the others: the eye, the olfactory mucous membrane (smell) and the lateral line (orientation). Selachians, such as sharks, have the undeserved fame of being shortsighted and weakeyed; we shall have a good deal to say about this later. But it is perfectly true that they do have an excellent sense of smell, and a third very important organ is the lateral line which, as we shall see in the next chapter, gives the hunter precious information.

In a conventional fish, these three sense organs are on the dorsal, or top side of the head, with the nostrils at the end of the nose.

THE DOMINANT SENSE ORGANS BECOME SPECIALIZED

When the selachian's nostrils and lateral line shifted ventrally along with the mouth, they became specialized in collecting information about what was happening below the animal. With conventional

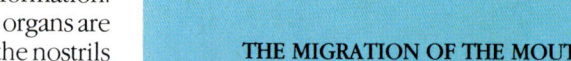

hypothesis of the lengthening of the head

hypothesis of the mouth rotating ventrally

THE MIGRATION OF THE MOUTH

In all fish, old and present-day, and in the ancestors of the selachians, the mouth was at the end of the nose, the most suitable place for collecting food, preying and so on.

But as the selachians evolved, the mouth moved down ventrally, toward the stomach. This may have happened with some kind of rotation, or because the head grew in length and developed that sort of pointed snout that most sharks have and that helps them a lot with their swimming.

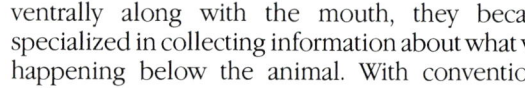

The ventral migration of the nostrils, toward the stomach, a consequence of the mouth rotating downward, enabled sharks to assess what lay below them as well. There was no more overlapping of eyes and nostrils, so each collected more specialized information.

The hammerhead shark approaches its prey in a spiral. Top left: the head of the hammerhead shark seen from below, with the eyes and nostrils set wide apart. Top right: an enlargement of a shark's olfactory pit.

positioning, the field of view partly overlaps the field explored by the sense of smell. When the nostrils and lateral line migrated, the eyes were left to explore the sides and back of the animal and the field below it was entrusted to the sense of smell and part of the lateral line. The information collected thus became more accurate, more specialized, and this gave advantages that well outweigh the disadvantage of the strange, uncomfortable position of the mouth.

THE SENSE OF SMELL

The structure of the shark's nostril is very complicated. It has an opening to let water in and another to let it out. So there is a continuous flow of water through the chambers where the close-packed laminae, or plates, of the olfactory mucous membrane collect even the tiniest pieces of smell information. Sharks are quite capable of recognizing a substance that is diluted one part to a million.

HUNTING WITH THE SENSE OF SMELL

With such a fine nose, we might say, the shark not only singles out its prey but also directs itself toward it. Our own nostrils are very close together; the tissues that identify smells are not very well developed. In sharks, on the other hand, the olfactory pits are wide apart, and their exceptional sensitivity enables the fish to assess and differentiate information accurately. If the prey is on the right, its smell will be stronger on that side and the shark sets off in that direction. If the prey is in the middle, both nostrils will supply the same information, so the shark goes off in a straight line, but this is quite rare.

THE TYPICAL HUNTING STRATEGY OF THE HAMMERHEAD SHARK

Sharks normally approach their prey in a spiral, because they tend to swim toward the side where the prey is and constantly compare smell information coming in from one nostril with information from the other to assess differences in intensity. And the technique is all the more advantageous the wider the nostrils are set apart, because the more independent they are, the more sensitive the nostrils become to differences in smell.

It is probably this greater difference and efficiency that explains the strange shape of the hammerhead shark. The hammerhead shark is a keen hunter of rayfish. The protrusions on the side of the shark's head not only keep the eyes well apart but the nostrils too, and this increases their sensitivity and ability to guide the animal toward the unsuspecting rayfish half concealed in the sand.

13. THE MAGNIFICENT SENSE ORGANS OF THE SHARK: THE EYE AND THE LATERAL LINE

When we think of sense organs, we inevitably refer to the ones we possess, and we use them as a yardstick for evaluating others. In actual fact, the sensitivity of man is far inferior to that of animals. Fish in particular get to know about their environment through organs sensitive to information unknown and completely incomprehensible to us. Even though it is one of the oldest fish, the shark too has a very specialized set of sense organs.

THE EYE

We already know about its sense of smell, and its eye is just as efficient. But we have to take a second look at the shark's reputation for shortsightedness, because there is something about it that makes it quite unique. In vertebrates that lead daytime lives, the light-sensitive cells of the eye, called cones and rods, are isolated from each other by fringes of black pigment that absorb rays of light once these have stimulated the cone or rod. In vertebrates that lead nighttime lives, on the other hand, crystals around the sensitive cells reflect what weak light there is, allowing the cones and rods to be stimulated many times. Eyes with these crystals appear to be luminous in the dark. Cats, dogs and owls have them, for instance. Well, our shark has both possibilities; during the day, cones and rods isolated by fringes of pigment that withdraw, at night, to reveal the crystals nighttime creatures have. So the shark is the only vertebrate with an adaptable eye, for day or night vision.

SHARK AND RAYFISH FOOD FIT FOR A KING

Certain sharks and rayfish have excellent meat and are delicious to eat: tope, dogfish, porbeagle and so on. The funny thing is that practically everywhere in the world, sharkmeat is sold under the name of some other, better-known or more prized fish. Mackerel shark, once decapitated, is sold as swordfish or even salmon, sometimes. Isurus, or mako shark, and the hammerhead shark are sold as tuna. In some parts of the world, rock salmon is the name given to a fish that is neither a salmon nor a rock-dweller; it is a *Scylliorhinida*, or dogfish, from a family of selachians that lives in the open sea. In Central and Northern Europe, slices of smoked shark are sold under the name of lake eel, while smoked rayfish becomes lake trout.

THE LATERAL LINE

But the shark's most interesting and mysterious sense organ — and all other sea creatures have it — is the lateral line and its offshoots. If we take a close look at the head and front part of the body of a fish, we see, on both sides, a series of holes in the skin, laid out in a row, one behind the other. They form a pattern that is now straight, as in the trunk and tail, now with elegant convolutions, as around the eye and near the mouth and nostrils. If we then take an even closer look at these holes, we see that they all comunicate with a channel that has groups of sensory cells in it. Sea water enters this channel and brings in information to the cells, which transmit it to the creature's brain. Now, what information is this? Recent studies of the function of the lateral line reveal it to be an extremely versatile sense organ. First of all, it helps the animal to hear the sounds of its environment; the sea creature only uses its ears to help it keep its balance, so sound waves are actually picked up by the cells of the lateral line. Then, a thrashing body, a wounded animal or an unaccomplished swimmer generates compression waves that are perceived by the shark's lateral line, and off he goes toward the potential prey.

The lateral line is also what is known as a thermoreceptor, since it can also measure minute water temperature changes, as small as a hundredth of a degree, even.

Further evolution of the lateral line enabled the shark to evaluate variations in the electrical field generated by its own muscles. But we shall have a better look at this when we come to talk about the shark's cousin, the rayfish, which has a very refined, sophisticated version of this sense.

TASTE

The shark's sense of taste is also probably very sensitive, but not much research has been done on it. What is certain is the presence of taste buds covering the whole of the animal's skin, from snout to tail, compact on the head, less so on the trunk.

The shark may also navigate using the Earth's magnetic field, because some of them periodically migrate to very precise destinations. But we know very little about this.

From what we have said, it is obvious how much information — completely unknown to us — the shark is able to gather about his environment. He knows perfectly well about everything happening around him, far or near, so that he is always ready to take advantage of any event that interests him.

Compared to the shark, our senses are very feeble indeed.

As we saw with the sense of smell, the lateral line is a powerful instrument for the hunting shark. The shark perceives that the compression waves of a prey moving to his right in the water are stronger on his right lateral line than on his left. He assesses the difference between the two and continuously corrects his route, following a typically spiral movement.

Right: the skin of a fish, showing the structure of the lateral line. One other function of this organ is to send messages to the shark's brain about the sound waves propagated through the water, because the fish uses its ears only to help it keep its balance.

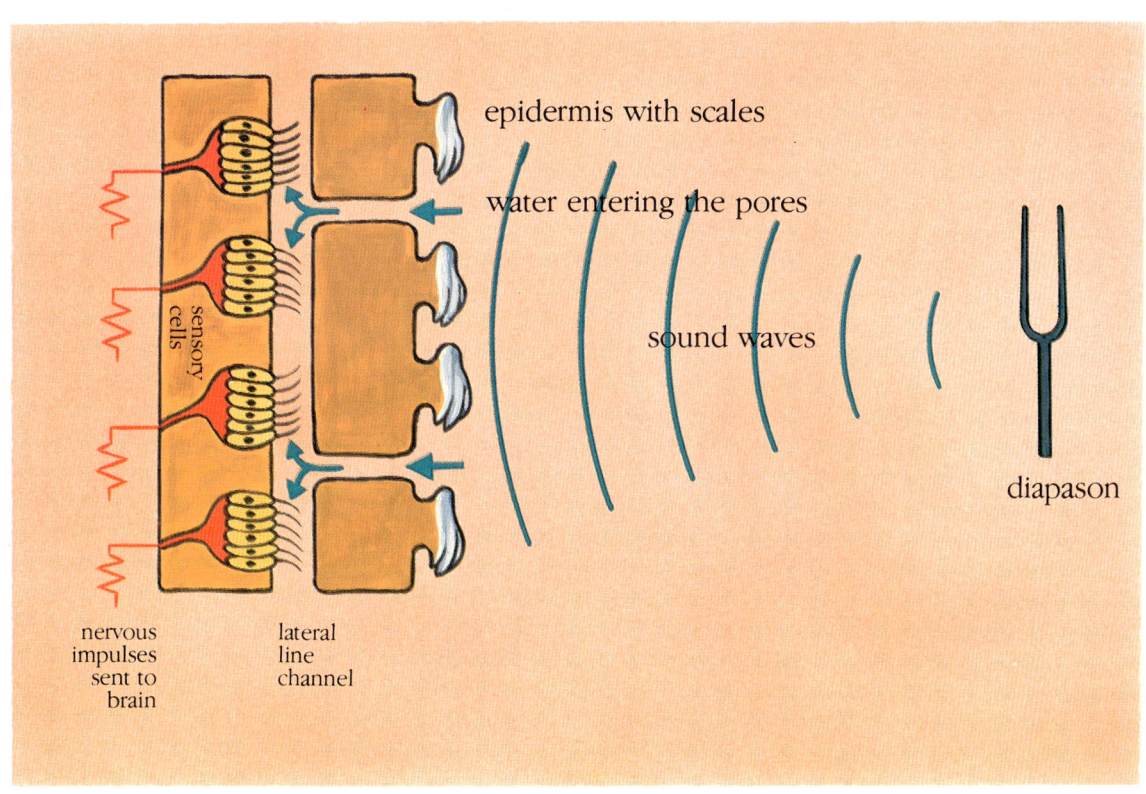

Left: the lateral line (dotted) in cartilaginous fish goes over the head, on both cheeks and chin, and down both sides to the tail. This sense organ, which communicates with the outside world by means of pores, has many functions. For instance, it perceives undulating or linear vibrations produced by a fish, be it moving or still.

The insert shows the taste buds that cover the shark's back. They are enclosed in little water-filled chambers that communicate with the outside.

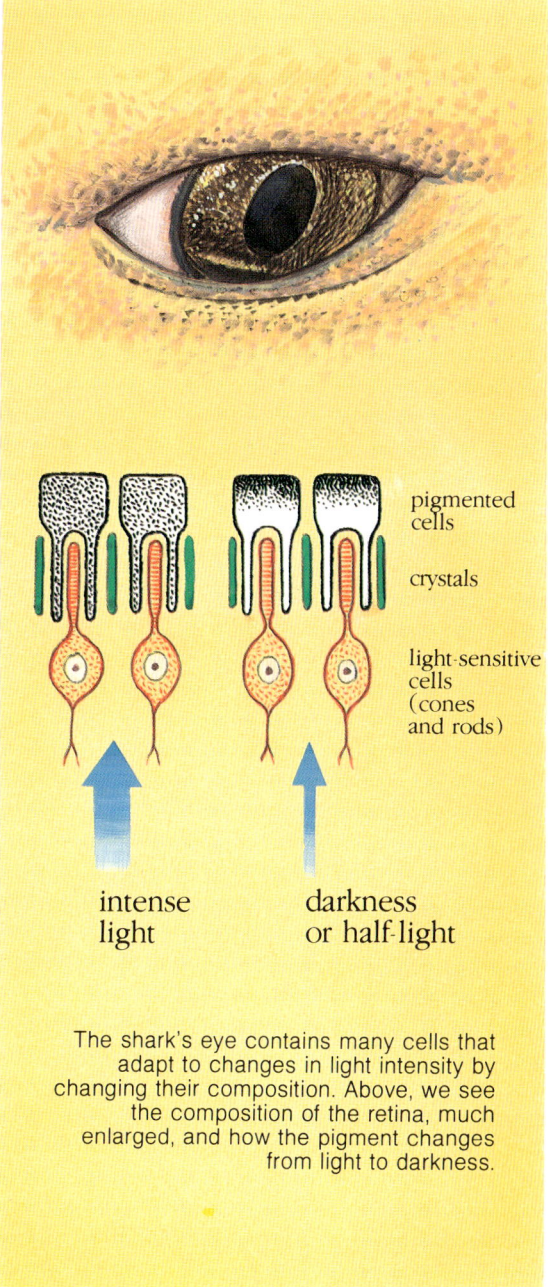

The shark's eye contains many cells that adapt to changes in light intensity by changing their composition. Above, we see the composition of the retina, much enlarged, and how the pigment changes from light to darkness.

14. WHAT SHARKS EAT AND HOW

Whatever food there may be in the sea, no matter how large or small, there is always a shark there to gobble it up. Apart from the ferocious sharks of fact and fiction, we also have quite innocent sharks that would never harm man, that roam the ocean depths using their fine sense of smell to find food.

FILTERING SHARKS

The largest living shark is the whale shark, which can be as long as 60 feet and weigh 15,000 pounds. It feeds on little fish and shrimp that it catches by filtering water just as the whale does, but a little differently. The liver of this strange shark is gigantic and very rich in oil — it can contain up to 130 gallons of oil — that keeps it afloat and upright, with its mouth open, and almost touching the surface. It swims in this position, so the water entering its mouth and leaving through its gills is filtered and all the little organisms in it are retained.

PREYING SHARKS

But when we talk of sharks, we usually think of the great hunters, like the white shark (which is really gray), the blue shark or the tiger shark. The white shark is by far the most dangerous, because it lives near the coasts, where it likes the food and it can and does settle on the bottom to rest. In Australia, some unfortunate swimmer is always treading on a shark sleeping comfortably not far off the beach. But the great hunters are nearly always way out and have little chance of meeting man. They are incredibly greedy and eat everything they find, without distinction. In the stomach of one white shark, they have found a ham, a number of sheeps' feet, the back half of a pig, the head and paws of a dog, strips of a horse and a few canvas bags. There must have been a shipwreck somewhere.

The insatiable appetite of these sharks can be explained. First and foremost, they are generally fish that can never stop swimming; as we have seen, their specific weight would drag them down into the abyss if they stood still, so to speak, so they need a good deal of food to keep on endlessly swimming without tiring. Then, their intestine, primitive and not very practical, is straight; it is not twisted and curved as it is in the more evolved vertebrates, so the route of its food is short and the digestive juices never have enough time to dissolve food completely. And the surface area of the intestine is so small that it absorbs the products of digestion very badly. Even though there is one spiral fold in the intestine to increase the absorbent surface, it does not work very well. So the shark never fully digests its prey. Hence its insatiable appetite.

THE POSITION OF THE MOUTH

To hunt and tear off strips of flesh, the shark has a mouth armed with formidable teeth. However, the position of its mouth is ridiculous, making things very awkward for the animal. But we have heard of the advantages of a ventrally positioned mouth and sense organs, so the inconvenience is quite acceptable.

With its ventral mouth, a shark has to attack a more voluminous prey than himself belly first, for only in this position can he sink his teeth into the victim, bite strongly and tear the flesh away. In fact, he has to shake his head quite vigorously to do this.

Indeed, one noted expert on sharks found out at his own expense how baby bull sharks have already learned to use their teeth while still inside their mothers' bodies. He opened the wall of the uterus of a female caught just a short time before, slipped his hand into the opening to help the babies out and was badly bitten.

Right-hand page: apart from the classic shark as we all know him, there are quite harmless forms, like the enormous whale shark, that can be approached without risk, and the nursehound, or small spotted dogfish, which roams the bottom looking for food.

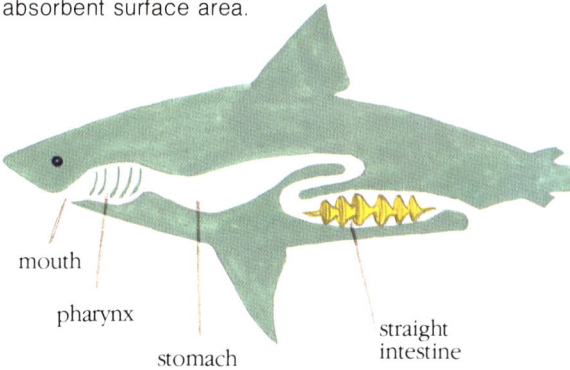

With its ventral mouth, the shark has to take up a special position to seize a prey larger than itself.

The intestine of the shark is in a straight line, with one fold in it and a spiral valve that increases the absorbent surface area.

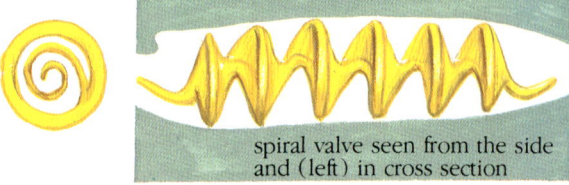

mouth

pharynx

stomach

straight intestine

spiral valve seen from the side and (left) in cross section

TEETH

The teeth of hunter sharks are truly powerful and highly efficient. Teeth are generally triangular plates with saw edges. They are only loosely attached to the upper and lower jaws, so they come out easily. But you will never see a toothless shark; as soon as a tooth comes out, another is immediately ready to take its place. Each tooth has a whole row of others in size order, ready to take its place. Sharks are the only creatures to have this kind of system. It may seem somewhat fragile, but it is really efficient.

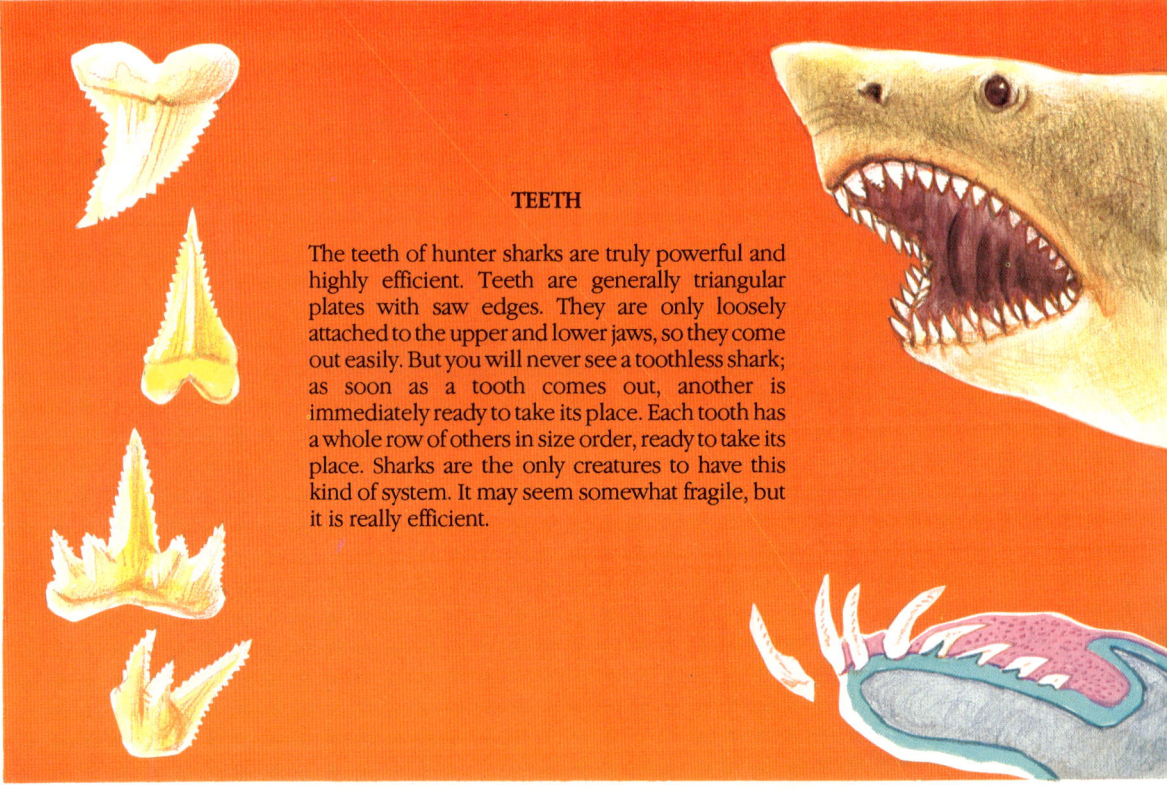

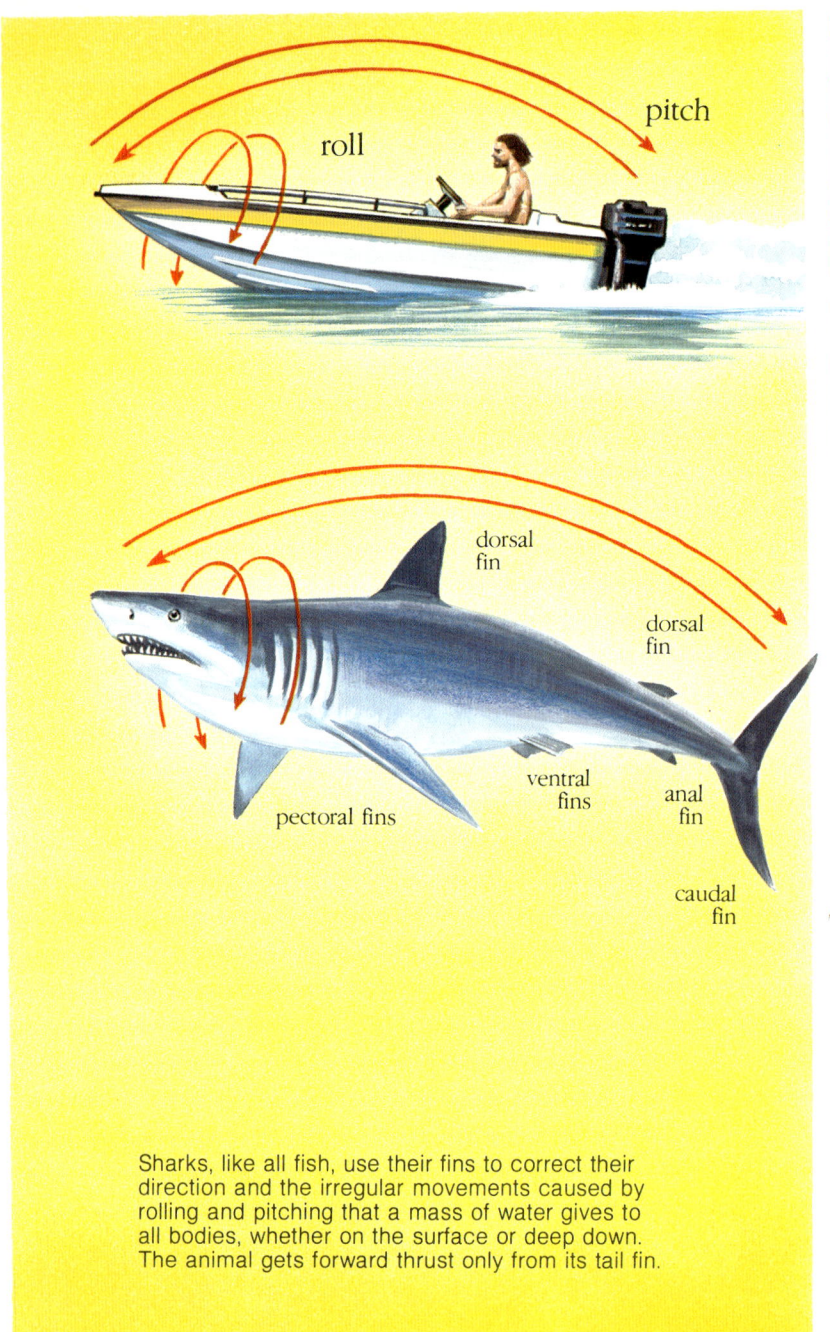

Sharks, like all fish, use their fins to correct their direction and the irregular movements caused by rolling and pitching that a mass of water gives to all bodies, whether on the surface or deep down. The animal gets forward thrust only from its tail fin.

15. HOW SHARKS SWIM AND HUNT

Sharks have the bodies typical of fish. Their streamlined shape lets them glide through water without difficulty and the powerful muscles of their tails give them good forward movement. This is the same with all fish, and also with the ichthyosaurs, the reptiles that lived in the Secondary Era, and our present-day members of the whale family, the cetaceans.

THE FUNCTION OF FINS

All these forms of vertebrates that adapted to life in water also had fins: single ones on the back and underside and pairs on the sides of the head and the front part of the body.

Fins have extremely important functions as regards the irregular movements that a body would normally perform in water. We all know about the rolling and pitching of a boat, which sometimes makes you feel a little dizzy and ill. We cannot go so far as to say that sea creatures suffer from seasickness, but we do know that fins are used to automatically correct or at least reduce rolling and pitching movements. And the front fins have another function concerning movement; when the tail provides forward thrust, the creature needs some means of braking and reversing its body. In teleosts, or bony fish, the front fins can be moved and used as brakes and, as we might say, as a reverse gear.

Funnily enough, sharks cannot move their front fins as the teleosts do; they can only use them obliquely to provide the upward thrust that helps to stop their high specific weight from dragging them down to be bottom.

This is why the shark is perhaps the only fish that cannot stop.

All it can do is stop providing forward thrust with its tail and wait for its body to come to a standstill as a result of friction with the water. And sharks cannot swim backwards.

JET SHARKS

Certain strong-swimming sharks use not only their tails but also their gills to provide themselves with

Shark

Pilot fish

Sucker fish

Sucker fish

Large sharks often have company in the form of pilot fish that swim in front of them or sucker fish that stick themselves to the shark's skin like suckers. Popular belief has it that these fish guide the shark toward its prey but, in actual fact, they offer him nothing of use in return for the scraps from his banquets.

The hunting technique of the thresher shark is rather peculiar. When he spots a school of fish, he begins to swim around it in an ever-tighter circle, terrifying the fish by threshing his long tail wildly. Once they are all bunched together, the shark comes in for the kill.

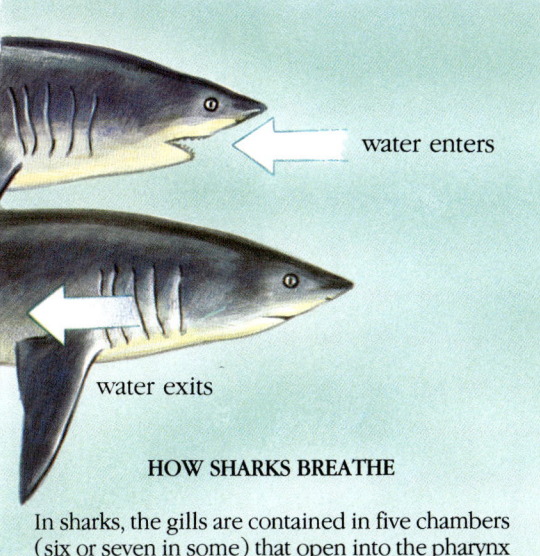

water enters

water exits

HOW SHARKS BREATHE

In sharks, the gills are contained in five chambers (six or seven in some) that open into the pharynx at one end, and into the sides of the head at the other. To circulate water, the shark opens its mouth wide while keeping the gill slits closed. It then opens the gill slits while closing its mouth to force the water to pass from the pharynx into the chambers and then to the outside.

forward movement! While they are swimming, their mouths fill up with water that they drive out of the gill slits at high pressure, thus helping the tail in providing forward thrust. Some sharks do not even bother to keep opening and closing their mouths to do this; they swim with their mouths permanently open.

THE TECHNIQUE OF THE THRESHER SHARK

We have already seen how sharks hunt. But the thresher shark deserves special mention. He owes his name to the presence of a long caudal, or tail, fin that some have imaginatively found similar to the tail of a fox. Hence the creature's Latin name is *Alopias vulpinus; vulpinus* means fox. This fish has an unusual way of hunting. It lives off schools of little fish but does not bother to catch just one at a time. It swims around the shoal in ever narrower spirals threshing its long tail wildly (hence the name most people give it). Bewildered and terrified, the little fish all huddle together chaotically in the center of the spiral and then, unable to flee, make an easy prey for the ever-hungry shark.

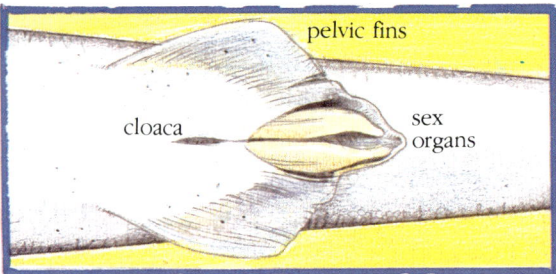

The difference between the two sexes. Above, the male; below, the female. The sharks are seen from the bottom, ventrally, at the level of the pelvic fins.

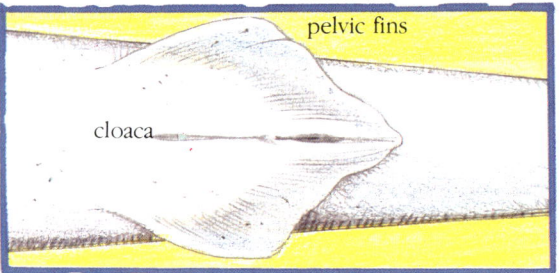

16. HOW SHARKS REPRODUCE

FERTILIZATION

One of the many contradictions in the shark's anatomy, which has primitive features alongside more evolved ones, is the presence of sex organs that allow internal fertilization. Laying eggs and spermatozoa in an external environment and letting fate decide whether the two meet or not is an archaic characteristic. Only in more evolved forms does internal fertilization appear, with sex organs to help the eggs and spermatozoon to meet. Although our male shark is a very old fish, he has sex organs called claspers at the sides of the cloaca — the genital, urinary and fecal aperture — that are the result of a modification of a ray and the pelvic fins. Indeed, the easiest way to tell a male from a female shark is just to look for his two sex organs.

OVIPAROUS SHARKS

There are also certain conflicting ways in which the eggs develop. One group of sharks, called oviparous (egg-born), lays eggs that normally become attached to plants on the sea bed. The embryo develops inside the egg, and has a bulky yolk sac for company; rich in food, the sac provides the embryo with nourishment throughout its development. This is a very ancient, primitive method of reproduction; the egg and the embryo are left to their fate, and there is no lack of enemies and dangers around them.

VIVIPAROUS SHARKS

But there is another group, the viviparous (born alive) sharks, that develops in a far more modern, efficient way. The eggs are not laid, but kept in the mother's oviduct. In this case, the yolk is not only used to nourish the embryo, but also to set up relations with the mother's vessels and make exchanges of food possible between embryo and mother. In this group, the embryo develops inside the mother's body and receives food from it. It is an extremely modern form of development, very similar to that of mammals, man included. The embryo stays in the uterus, which is no more than a specialized section of the oviduct, and relates with its mother, who feeds it and rids its metabolism of waste matter throughout the pregnancy. So there are strong similarities in the development of viviparous sharks and mammals. And, of course, just like the mammal, the mother shark gives birth to a perfectly formed baby shark at the end of her pregnancy; it leaves the cloaca head first and the rest of the body follows. This is a much safer way of developing than just abandoning the eggs in a hostile environment. And when the newborn shark becomes free, it is already quite capable of self-defense.

"Oviparous" is the term used for sharks that lay their eggs in an environment outside the female body. The eggs we see above are like capsules covered with a horny, transparent shell that protects the embryos. At the corners are long filaments that wrap around the nearest weeds and keep the precious egg suspended.

THE YOLK SAC

If the vertebrate's egg is laid in an environment where there is no possibility of the mother feeding the embryo in any way, it needs a store of substances to help its development. The embryo has the yolk egg for this purpose. It is attached by a sort of umbilical cord full of vessels that convey food from the sac to the embryo.

Left: the birth of viviparous sharks greatly resembles the birth of mammals. The baby shark leaves its mother's body already perfectly formed, and even has an umbilical cord, the remains of the yolk sac.

In certain viviparous sharks, like the bull shark, ferocious cannibalism provides the embryos with food.

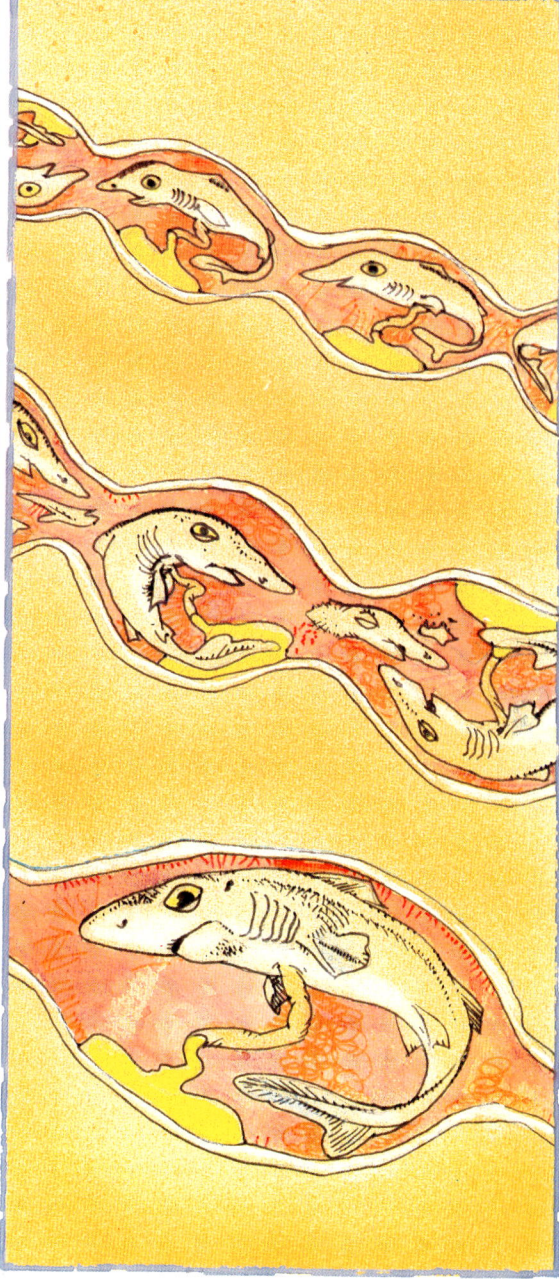

"Viviparous" is the term used for sharks that develop their eggs inside the female body, from which the egg also takes the nourishment it needs. Left: in the mother's oviduct, swollen like a uterus, we see the embryo of a viviparous shark; its yolk sac has become a kind of placenta (which enables nourishment to pass from mother to baby), similar to the uterus of a mammal, which we see on the right.

The female mackerel shark, or porbeagle, goes on producing eggs while her two embryos, one in each oviduct, are developing. These eggs are not wasted, however, because the little porbeagles, in the dark of their mother's inside, seek them, find them and gobble them up.

CANNIBALISM

What happens with the ferocious, extremely dangerous bull shark ably bears out its notorious reputation for aggression. The female produces a series of eggs that stop along the two oviducts. A whole line of embryos then forms, feeding off the yolk sacs at first. Once these have been devoured, a furious battle begins; the embryos fight it out and the victorious eat the conquered. The number of embryos gradually decreases until there is only one champion left in each oviduct, by which time the baby sharks are ready for birth. No wonder, then, at the tremendous greed of the bull shark, who has to survive such a ferocious process of selection even before it is born.

Basking shark

Megachasma pelagios

17. THE SURVIVORS: LIVING SHARKS

As we have seen, sharks went through a serious period of crisis because of the numerous marine reptiles in the seas of the Secondary Era, some 250 million years ago, but once their competitors had disappeared, they quickly made a comeback and regained their lost dominion. Today, sharks are the largest, greediest predators in the seas, their only real competitors being killer whales.

It would be equally useless and boring to list all the sharks living today. Normally, there are no great or spectacular differences among the various species. Sizes change, as does color and shape of fins, mouth, teeth and habits, but the basic structure of the shark is the same. Some do, however, deserve special mention.

THE *CARCHARODON*, OR WHITE SHARK

This is perhaps the most typical and best known of all sharks.

When we talk of a preying shark, or draw one, it is usually a *Carcharodon* (which means sharp-toothed). This fish attacks and devours everything it meets, sometimes even boats. It likes to follow ships, snapping up whatever is thrown overboard. The name "white" shark is definitely misleading; the color of the skin on its sides goes from gray to black and only the belly is white. It is exceptionally large for a preying shark, growing as long as 40 feet in length. But one of its direct relatives, the *Megalodon* (large-toothed), which became extinct a few million years ago, was an incredible 100 feet long. When you think that the fearful mouth of the white shark can swallow up a whole man without even scratching him, you can imagine how huge the *Megalodon's* mouth must have been. It is a good thing for us that it is extinct, for it could quite easily eat a whole boat — and all its occupants along with it — in one go.

THE *ISURUS*, OR MAKO SHARK

Some people call this fish the tuna shark because it is so delicious to eat. It is a tireless swimmer; day and night, it roams the seas looking for prey. It is perhaps the most dangerous shark for anyone swimming too far from the shore. Catching mako sharks is a real spectacle. They are fished from fast motorboats and, once hooked, thresh about frantically to get free, making huge leaps right out of the water.

THE *PRIONACE GLAUCA*, OR BLUE FISH, OR BLUE SHARK

This is another highly dangerous shark, on account of its appetite, but it does not grow very large. This fish also likes to follow ships for the leftovers rather than go hunting on its own steam. So it tends to live near the coastline, making itself a very dangerous threat to man.

THE *LAMNA NASUS*, OR PORBEAGLE

This is a very common shark, both in the sea and in the fishmarket. The name (*nasus* means nose) derives from the shark's very long snout, much

Mako shark

Blue shark

White shark

DEFENSE AGAINST SHARKS

Much research is carried out in a number of countries in an attempt to discourage sharks from attacking man. For instance, curtains of air bubbles have been made to rise from perforated pipes on the sea bed offshore from the beaches. But instead of dreading them, sharks tend to play with the air bubbles, carrouseling through them time and again. For the time being, no other methods have given better results.

longer than usual. It is better not to play around with this fish; ferociously aggressive, it can cause great harm. It is famous because of the exquisite taste of its meat (another of its names is mackerel shark). As soon as it is caught, it gives off a nasty smell that goes away, however, after a few hours. It is sold under its own name, and as swordfish and even as salmon.

THE *CETORHINUS MAXIMUS*, OR BASKING SHARK

Second in size only to the whale shark, which we have already met, the basking shark, some 40 feet long, normally swims along with its mouth wide open, filtering tons of water a day and collecting thousands of little fish in it. Neither of these giant sharks ever attack man. They are too busy eating little fish 24 hours a day.

THE *MEGACHASMA PELAGIOS*

Very few people know of this shark; only one of its kind has ever been caught, off the Hawaiian Islands. Nothing is known of its lifestyle. Nevertheless, it is interesting, because its mouth is at the end of its snout, just like the ancestors of the whole class.

It may well be a living representative of the sharks that flourished in the Devonian-Permian Period, that is, before the shark's mouth migrated ventrally. If this is true, the *Megachasma* is one of the oldest known living fish, a true and proper living fossil.

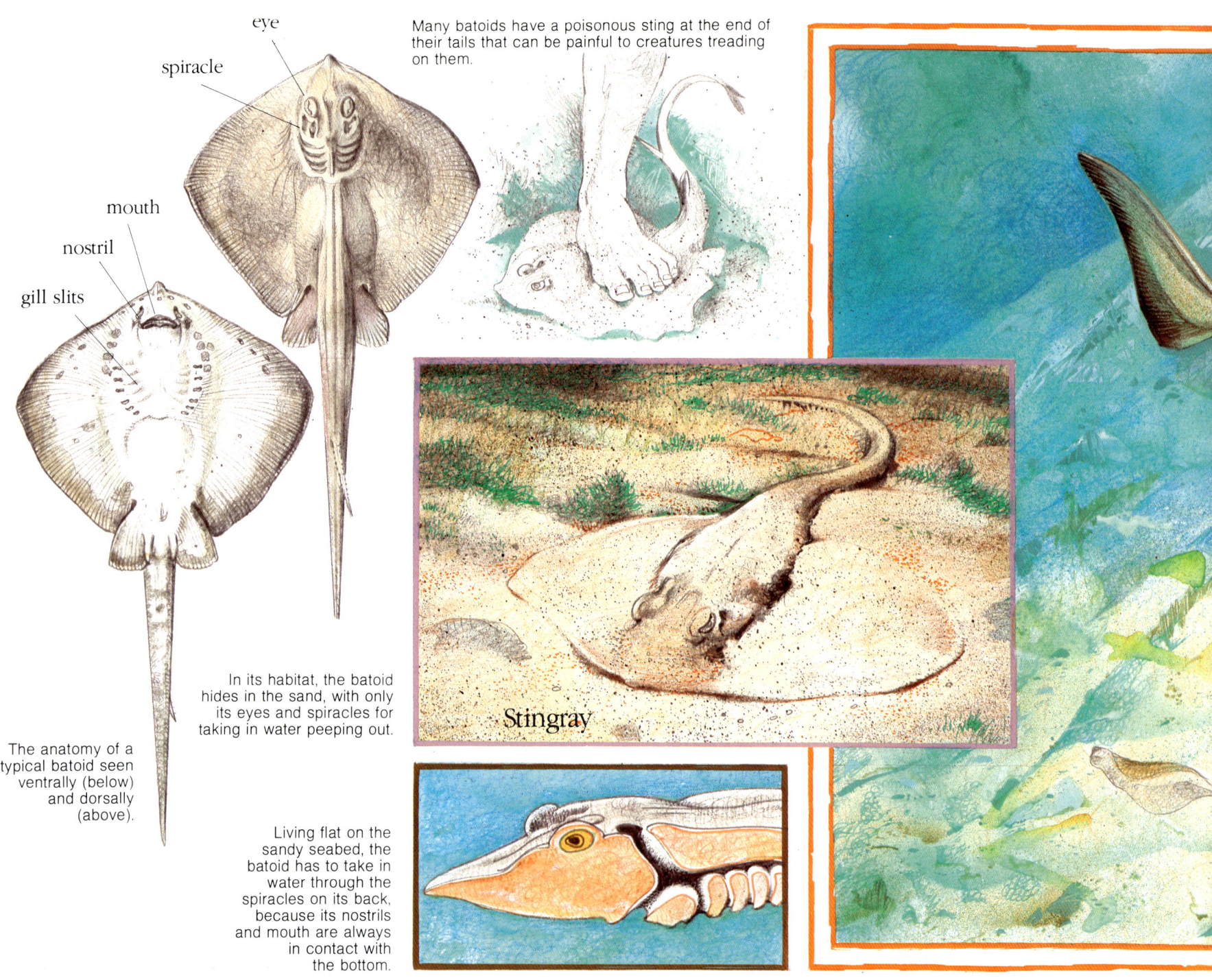

eye
spiracle
mouth
nostril
gill slits

Many batoids have a poisonous sting at the end of their tails that can be painful to creatures treading on them.

In its habitat, the batoid hides in the sand, with only its eyes and spiracles for taking in water peeping out.

The anatomy of a typical batoid seen ventrally (below) and dorsally (above).

Stingray

Living flat on the sandy seabed, the batoid has to take in water through the spiracles on its back, because its nostrils and mouth are always in contact with the bottom.

18. RAYFISH, OR BATOIDS

The batoids are strictly related to the shark family, because they too have cartilaginous skeletons. The differences lie in the flat body and the gill slits, which open ventrally, beneath the animal, and not on the sides, as in other fish, sharks included.

A CREATURE OF THE DEPTHS

Even though there are species of rayfish that swim rather well, the batoids evolved as creatures of the depths. Their flat bodies and special way of moving allow them to bury themselves in the sand, both to find food and to hide from enemies. But living buried in sand creates a serious breathing problem. If water were taken in through the mouth, as in most other fish, it would bring débris, sand and little oxygen with it. The first gill chamber, called the spiracle, is used to overcome the problem. Water is taken in through its openings on the back of the fish. In fact, when it is under the sand, the rayfish lets its two eyes and the two tubes of the spiracle peep through.

THEY FLY RATHER THAN SWIM

But batoids can and do swim. The eagle ray (*Myliobatis aquila*) is a magnificent swimmer, owing its name to the way it moves in water. In all fishes, the bending backbone, with its muscles and fins, provides the thrust necessary for movement. But batoids have rigid, practically unmovable backbones and their enormous lateral fins move up and down, just like birds. So we might say that, rather than swim, rayfish fly through the water. And the image of an eagle certainly does come to mind when you see the

HOW RAYFISH SWIM

When rayfish swim, their large, wing-like pectoral fins make undulating, up-and-down movements that wave along from head to trunk and tail. The technique, very similar to the one adopted by birds, is the consequence of the very flat front part of the body, which greatly restricts movement of the backbone.

The sawfish, a very curious batoid, uses its powerful weapon mainly to drive its prey out of the mud or sand. The eagle ray, with its long, pointed tail and large, triangular fins, glides over the seabed in search of shellfish for food.

majestically gliding eagle ray. Like other batoids, this one also has one feature that makes it dangerous; at the end of its long, pointed tail is a toothed sting that it uses for defense. The sting is particularly painful and sometimes mortal, being impregnated with a poisonous substance secreted by a gland.

FOOD

The batoids' food is rather monotonous; little animals collected on the seabed, like mollusks, shrimp and octopus. Their mouths have no razor-sharp teeth with which to tear or rip their prey.

Their teeth tend to be flat and can crush shells and hard skeletons. They seek their prey mostly by using their sense of smell, very refined and sensitive, while their little eyes, on their backs, are used only to keep a lookout for any predators lurking around.

THE SAWFISH

The batoid that arouses our curiosity most is the *Pristis pristis*, commonly called the sawfish. Its nose, or snout, rather, is exceptionally long and armed on both sides with sharp, pointed teeth. At first sight, with its streamlined body, the creature seems more like a shark than a rayfish, but the position of its gill slits classifies it as a batoid. As in all others in the same sub-class, they are on the underside of the fish. Sawfish grow to considerable dimensions, sometimes reaching 35 feet in length. Meeting an animal of this size is always an unforgettable experience, and not without danger. Its formidable rostrum, as the saw-nose is called, is used to hunt prey. Swimming around near the seabed, the animal prods the sand or slime with its "saw," which snatches the unfortunate victims from the bottom and tears them to pieces

before the banquet begins. But the sawfish also uses his weapon to defend himself from predators like sharks or, if he feels threatened, against over-curious, trespassing underwater swimmers and divers. The wounds he inflicts are not very deep, but flesh tends to be ripped away, and the victim can bleed to death.

One last curiosity about the sawfish is the birth and development of its young. *Pristis* is, in fact, a viviparous fish; that is, the embryo develops inside its mother's oviduct and is fully formed and efficient at birth, except for its saw, which remains soft and useless during development, covered by a sheath-like case. So the mother runs no risk during the development of the young sawfish, or during labor. Only when the newborn fish begins to swim freely in the sea do calcium salts quickly strengthen the saw and provide it with the teeth that will make it so strong and deadly.

19. THE ELECTRICAL ORGANS OF CERTAIN BATOIDS

When it discharges, the electrical organ creates electric fields that foreign bodies disturb. The electric ray perceives these changes and gets to know his surroundings in every detail.

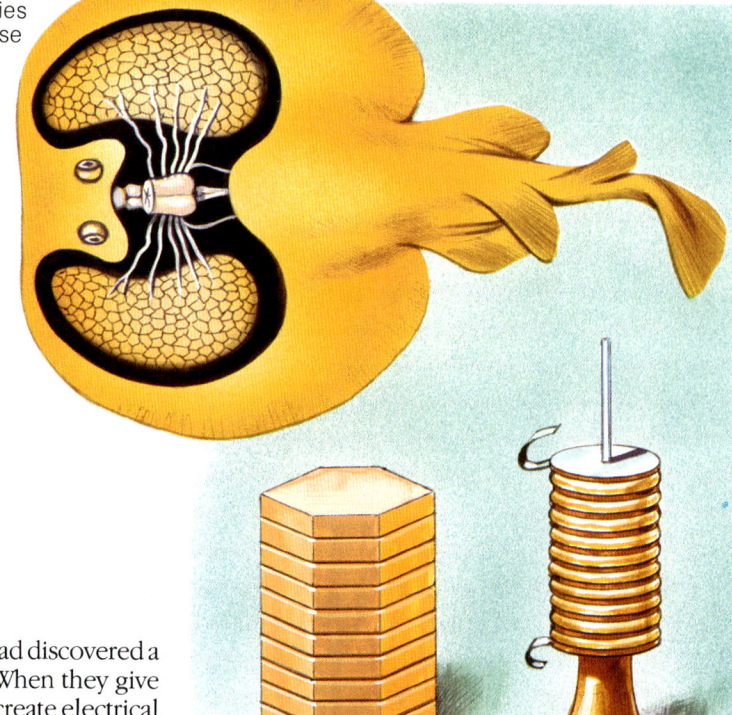

At the base of their pectoral fins, electric rays have powerful electrical organs (formed of atrophied, greatly modified muscle tissue) that give off violent shocks. At right, we see a section of the electric ray from above, showing the electrical organs. Top opposite page: the deadly spotted eagle ray, to be avoided at all cost.

When, in ancient times, a fisherman accidentally touched an electric ray, he felt an intense stimulation, shivered violently and briefly and was stunned. The earliest naturalists, like the Greek, Aristotle, and the Roman, Pliny, looked into this, tried to explain it, but could conclude no more than that these animals possessed some "evil fluid." And that explained nothing. But we know that these great researchers could not go further, because they did not know about electricity. Research into electricity only began in the 17th century.

The discovery of the battery and all its possibilities meant that studies on the electricity in animals could be undertaken. It was soon seen that the "evil fluids" certain fish — and not only rayfish — possessed were none other than real electric shocks.

THE ELECTRICAL ORGAN OF THE ELECTRIC RAY

Rayfish, electric rays especially, can give off discharges of electricity as powerful as 200 volts, enough to light a bulb. How do they manage to do this? If you are very careful and make sure the electric ray is dead before you approach it, you will observe two jelly-like masses on its sides, at the base of its fins. They are called "electrical organs" and produce electricity by exploiting a special property of the muscles.

When a muscle contracts, it changes the magnetic field around it because two poles are created, the same as in a battery.

The electrical organ of the electric ray consists of a series of small, flat hexagonal muscles in vertical columns.

These muscles have lost their ability to contract but have retained their ability to create discharges of electricity. So they behave like a lot of small flat batteries when the animal is resting and are all charged immediately when the electric ray sends them a message through its nerves, telling them to produce an electric shock. Since these "batteries" —electric plates would be a better term — are arranged one on top of the other in a series, as in an electric torch, their voltages are all added together, which explains why they can discarge up to, and more than, 200 volts. Being formed of many groups of electric plates, the electrical organ can develop a very high, dangerous current.

THE FUNCTION OF THE ELECTRICAL ORGAN

The main function of the electrical organ is to stun outsiders and enemies and to kill small prey. This attack-defense task was instantly understood by the early researchers, but only after they had discovered a second, no less important function. When they give off electrical discharges, electric rays create electrical fields, the lines of force of which surround their bodies and become ever weaker toward the outside. The extraordinary fact is that the batoid "feels" these lines of force and perceives all changes in them through the lateral line that we have already seen. In fact, if some animal ventures into this field, it causes distortion of the lines of force that the rayfish notices instantly. Also, from the intensity of the disturbance, it perceives the kind of intruder and its size and so knows exactly what attitude to adopt. So apart from being a powerful means of attack or self-defense, the electrical organ is also an exceptional sense organ, one that can tell the batoid all about its surroundings, with all the variations that take place.

The electrical organ is not exclusive to the batoids; it formed independently in other groups of fish, the teleosts, for example, and especially in those groups that live in dark or murky waters and need some help to make up for their inefficient eyes.

THE BATTERY

The first battery was conceived in 1800 by Alessandro Volta. It comprised a stack of alternating copper and zinc disks lying one on top of the other, separated by acid-soaked fabric. Each element created a flow of electricity that was added to the flow created by its neigbor and so on until a strong current was accumulated at the end of the battery. Thus was "invented" an instrument for producing electricity that fish had already invented millions of years before.

Here, we see the similarity between Volta's battery (right) and the schematized crossection of the tissue that produces electricity in the electric ray (left).

Spotted eagle ray

Chimaera

THE *CHIMAERA*, OR RABBIT FISH

There is a small group of fish with cartilaginous skeletons known only to the specialists because it is represented by only one living order, the sole survivor of a much broader group that lived in the seas from the Devonian to the Jurassic Periods, 395 to 135 million years ago. The name given to this order, holocephalians (all-head), is not very kind, but does stress the extraordinary size of their heads. And if that were not enough, the animal's specific name is *Chimaera monstrosa* (monster she-goat). Doubtless, the fish was not deeply loved by the scientists that first described and baptized it.

Holocephalians grow quite large, three to four feet in length, but they are harmless; their mouths are relatively small and are used only for seeking food on the sandy or muddy seabed.

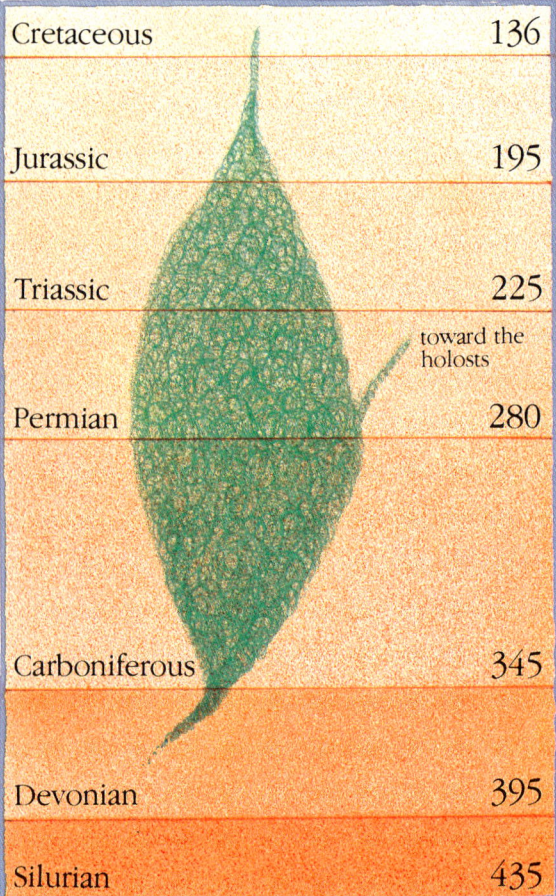

Cretaceous	136
Jurassic	195
Triassic	225
	toward the holosts
Permian	280
Carboniferous	345
Devonian	395
Silurian	435

The expansion of the chondrosts over the periods, in millions of years.

20. BONY FISH: THE CHONDROSTS

THE ACANTHODIANS, ANCESTORS OF BONY FISH

More than 400 million years ago, in the fresh waters of the Silurian Period, little fish bristling with spikes, the acanthodians, spent their time swimming around, ignorant of just how important they were to be in the history of fish. Much work is done on the acanthodians nowadays, and much controversy exists. Some say they are older than the placoderms and, hence, the precursors of all fish; others, the majority, say that they were an early offshoot of the placoderms and the starting point for all bony fish. We shall accept the latter hypothsesis, according to which the history of fish begins with the appearance of the great group of placoderms, all destined to become extinct, but from which two phyla of fish developed, autonomously and independently : fish with cartilaginous skeletons, the chondrichthyes (the future sharks, as we have seen) and fish with bony skeletons, the osteichthyes, initially represented by the acanthodians. The new fish spread rapidly throughout the Eart's fresh waters and, only a few million years after their appearance, they originated two phyla of bony fish, completely different in structure and in the destiny awaiting them.

FISH THAT BREATHE AIR

One group, called *Sarcopterygii* (flesh-winged), colonized the boundaries between water and air of the warm, fresh waters of the Devonian and

The bichir, the sturgeon and the paddlefish are among the few living representatives of the chondrosts group.

Carboniferous Periods. As we shall see in the last chapter or, better, in the volume on amphibians, this environment obliged creatures to adopt quite special anatomical characteristics, such as the ability to breathe the oxygen in the air and to move on stiff fins.

THE MORE TYPICAL FISH

Another phylum, the *Actinopterygii* (ray-winged), maintained and perfected the typical anatomical organization of fish destined to live in water. The long path of this line of evolution, however, was neither continuous nor progressive; it went along in fits and starts.

THE CHONDROSTS

The first great group of *Actinopterygii* we meet does not yet have a true bony skeleton. The fish have primitive features, such as bony scales on their heads, straight intestines and so on. The many forms of this group, or super-order, called chondrosts, retained dominion over their waters for a long time, from the Devonian to the Triassic Periods, some 150 million years. They probably originated in fresh waters but later emigrated to the seas as well, where they joined the sharks to become the second main type of fish of the period.

The fortunes of the chondrosts began to fade when a new type of bony fish appeared, more modern, with better bony structures.

They were the holosts, and we shall meet them in the next chapter. The presence of more efficient fish plunged the chondrosts into a state of crisis; they began to die out and finally became extinct at the beginning of the Jurassic Period. Only two groups have survived to our times: the polypterids (many wings) and the acipenserids (sturgeon-like).

THE POLYPTERIDS

The oldest of the two groups that survived are the polypterids. The bichir, for instance, has a snake-like body, is three to four feet long and has lots of little

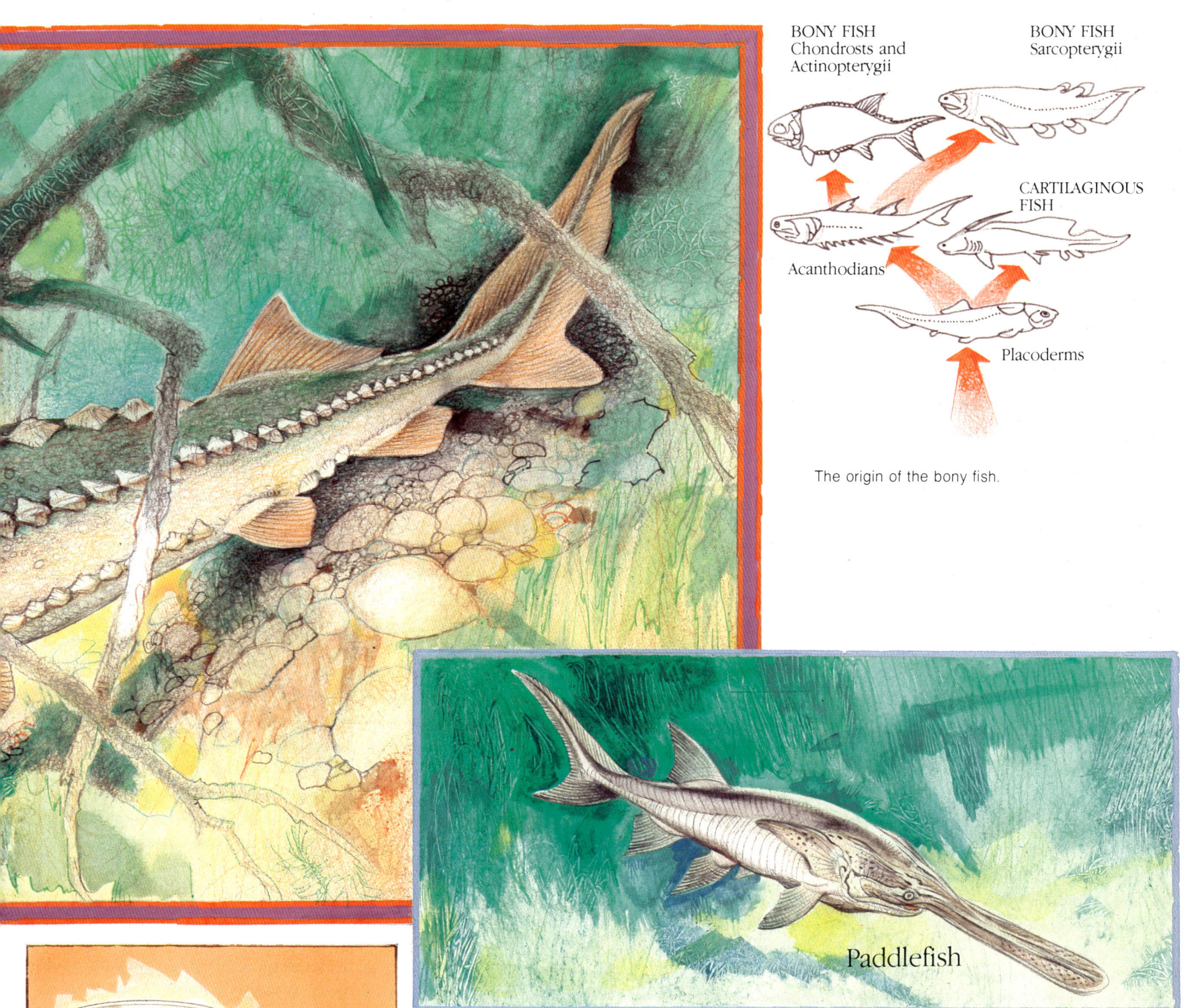

The origin of the bony fish.

CAVIAR

Caviar is made by salting sturgeons' eggs and applying a little borax to them. Each female produces about 10 percent of her own weight in eggs and, since these animals usually weigh anything up to 200 pounds, each fish yields some 20 to 45 pounds of eggs. There are as many types of sturgeon as there are of caviar. The absolute best comes from the sterlet, followed by sevruga and then beluga. Other types are decidedly inferior and some do not come from the sturgeon at all.

dorsal fins, while the front fins help the fish support its body and move around on the seabed. It is a predatory fish that inhabits the fresh waters of African rivers and lakes. It can breathe oxygen from the air through an air bladder and crawl out of the water when the river dries up to seek some deeper pool.

THE ACIPENSERIDS: THE STURGEON

The sturgeon too, well known for the deliciousness of its meat and even more for the exquisite caviar it produces, is a living fossil, a contemporary of the great group of chondrosts. It has many archaic features, such as a cartilaginous skull with a spiral valve and bony shield on the head, back and sides. The sturgeon is the largest freshwater fish; the record goes to one huge fellow that was 28 feet long.

Sturgeon are born in fresh waters and, when aged about 12 months, set out for the seas, where they complete their development. On reaching sexual maturity, after about 12 years, they feel the urge to reproduce, leave the sea and make their way back to the rivers where they were born. At this point, man steps in, greedy for the sturgeon's flesh and eggs; sturgeon are fished when the females, full of eggs, are swimming upstream again.

There is still an abundance of sturgeon in the Caspian and Black Seas and the rivers that flow into them, because they are very well protected, seeing how important they are from an economic point of view. They are declining, however, in the Mediterranean and seem to have died out completely in North America. But a cousin of the sturgeon does live on in North America; the *Polyodon spathula*, called paddlefish on account of the shape of its mouth. It produces a kind of caviar that has nothing in common with the real thing.

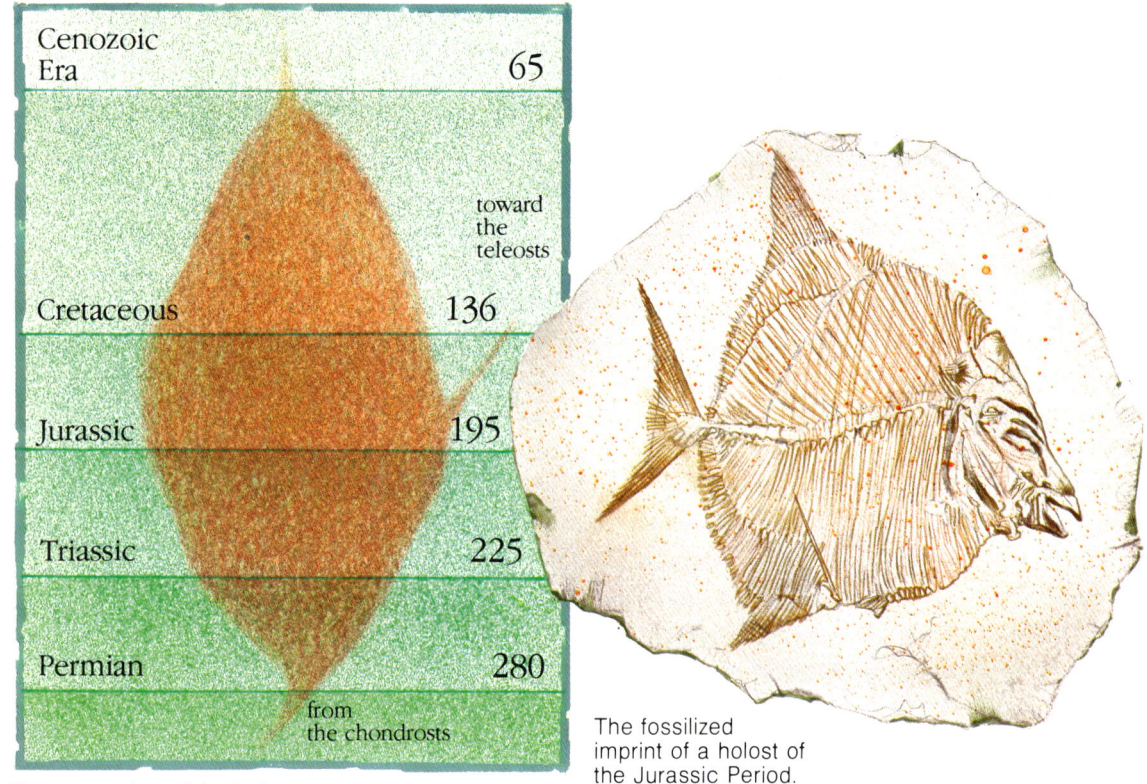

The expansion of the holosts over the periods, in millions of years.

The fossilized imprint of a holost of the Jurassic Period.

21. THE HOLOSTS

The story of the evolution of bony fish repeats the classic succession of forms by competition. The decline of the chondrosts, who had been masters for a long time, began with the appearance of a second, more modern and more efficient bony fish, the holost. Better adapted to the environment — a winner, therefore — the holost gradually ousted the old master in waters everywhere. It appeared about 225 million years ago, probably originating from a side branch of the chondrosts, but with a better capacity for evolution. Its period of greatest development was between the Jurassic and the Cretaceous Periods. But it was right in the middle of the same period that a third group of bony fish, derived from the holosts, came to the fore. These were the teleosts, which we shall see in the next chapter. In exactly the same way as millions of years before, the presence of a new phylum of fish, with perfect bony structures, differentiated and efficient, thrust the former holost masters into a state of absolute crisis and kept them there until, one by one, they disappeared from the scene.

Luckily for us, however, not all the holosts succumbed. Some are still with us today: the *Lepisosteus*, or gar, and the *Amia*, or bowfin or mudfish.

THE SURVIVORS

The survivors, the *Lepisosteus* (scaly-bone) and the *Amia* (needle), are North and Central American freshwater fish. Alongside rather advanced, evolved features, they retain certain primitive characteristics. The intestines, for instance, have convolutions, the same as all later vertebrates, but the air bladders of both fish are still used for breathing.

THE *LEPISOSTEUS*

Several species of gar live in the fresh waters of North and Central America. They are generally ruthless predators, capturing their prey with their long mouths armed with razor-sharp teeth.

One species, *Lepisosteus tristoechus*, reaches the respectable length of some 15 feet and, on account of the shape of its head, is called the freshwater shark, or the alligator-pike, but it is neither shark, nor pike, nor alligator. The anatomy of the gar is full of contradictions, very primitive in parts, but with other features that only came out with the amphibians and reptiles, like massive vertebrae and heads articulated with the backbone.

While no other fish, without exception, can turn their heads from right to left and so on, because their heads are rigidly fixed to their backbones, the *Lepisosteus* can, and indeed does, use the movement to help it catch its prey.

Gars do not like the cold and go into a kind of deep sleep in deep waters in winter while, in the summer, they love hunting around in the marshy ground, even in shallow waters poor in oxygen, because they can use their air bladders to breathe the oxygen in the atmosphere.

The *Lepisosteus osseus* lives in North American rivers, from Minnesota to the Gulf of Mexico. It differs from others of its species in the remarkable extension of its snout. Another of its names is, in fact, long-nosed pike.

THE *AMIA CALVA*

This fish lives in North America. It prefers shallow waters rich in vegetation where it lies low during the day and comes out to hunt at sundown. When oxygen and water begin to get scarce in summer, the bowfin can come up to the surface and breathe atmospheric oxygen. In the winter, on the other hand, it moves to deep water and falls into a deep sleep, waiting for the good weather to return once more. In winter, in the deeper points of rivers and marshes, you frequently

The male *Amia calva* takes his brood for swims. The young stay close together under his belly for protection. Along with the *Lepisosteus*, the *Amia* is one of the few living representatives of the holost group.

come across dozens of bowfins all hibernating together, side by side, something they would never do in the summer, when the males above all are alert, defending their territory. The females are larger than the males, some 30 inches in length, while the males can be recognized by the fine black patch with a golden-yellow edge on the root of their tail.

THE MALE *AMIA*, AN AFFECTIONATE FATHER

With the *Amia*, we come across a phenomenon that we shall not see very often, even in other classes; care for offspring, that is, a parent's defense of its own children.

During the mating season, the male prepares the nest, hollowing out a small hole and freeing it of mud and débris. After this, he starts to look for and attract a female. This is not so easy, because there is only one female for every three males, so two will have to remain bachelors that year, and their work will be to no avail. Once the female is found, the lucky male performs a complicated rite to invite her to lay her eggs in his nest. He then fertilizes them. The female goes off immediately afterwards, abandoning the nest, leaving all the chores to the male. Indeed, for the whole period during which the young develop, he has to keep the nest clean and ward off enemies. When the eggs hatch, the fry all remain together, packed under their father's belly.

This relationship goes on quite a long time, and it is very close; whenever the father goes any distance from the nest, he takes his offspring with him. If, because of some sudden darting movement, he gets separated from them, the group stays right where it is, quite still, waiting for him to return. If, by mischance, the father does not come back, the fry, vainly waiting, make a very easy prey.

22. MODERN FISH: THE TELEOSTS

In the mid-Secondary Era, 140 million years ago, while holosts dominated all the Earth's waters, new fish came on the scene, with perfect bony structures, much more efficient than the holosts. They were the teleosts. The newcomers had their origins in the holosts, but certain details of their anatomy were different, like the large air bladder, which they did not use for breathing atmospheric air but, as we saw in the chapter on specific weight, as a means of keeping the specific weight of their bodies in balance. They had a new intestine; and the new features of their skeletons and skins so perfected

SOME FEATURES

Teleosts pressed forward into the most difficult, inhospitable waters. Some graylings have adapted to life in hot springs that reach temperatures as high as 46 degrees centigrade (115 degrees Fahrenheit) with very high salt levels, while others live at freezing point, partly thanks to a kind of antifreeze in their tissues and blood.

They vary a great deal in size too. The giant of the teleosts is Brazilian; the pirarucu (*Arapaima gigas*), which grows as long as 15 feet. On the opposite extreme, the tiniest is the *Mistichthys* of the Philippines, never more than half an inch long, a contender with the *Pandaka pygmaea* for the smallest vertebrate record.

Many are the tales told about the long life of the teleosts. They tell, for instance, of a pike that lived in captivity to the ripe old age of 267 years, and a carp that lived to see its 150th birthday. These stories are difficult to prove. Anyway, the countless dangers and conflicts that nature provides would make it difficult to live so long.

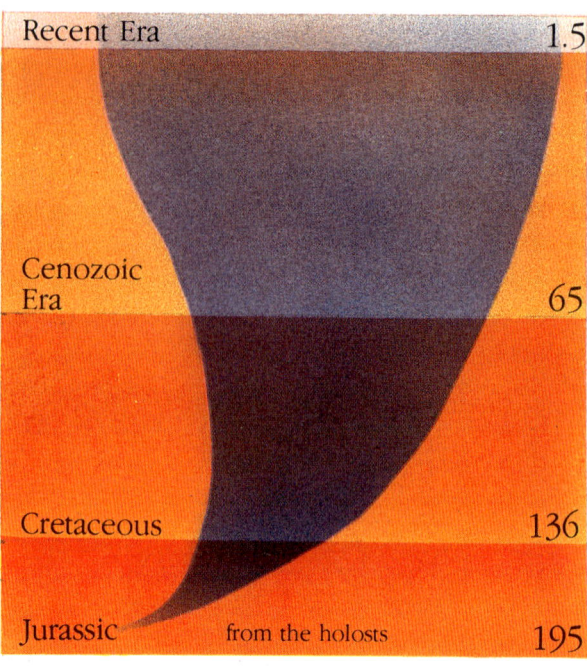
The expansion of the teleosts over the periods, in millions of years.

Certain teleosts, like the sea horse (or *Hippocampus*) and the porcupine fish have rather original, curious shapes.

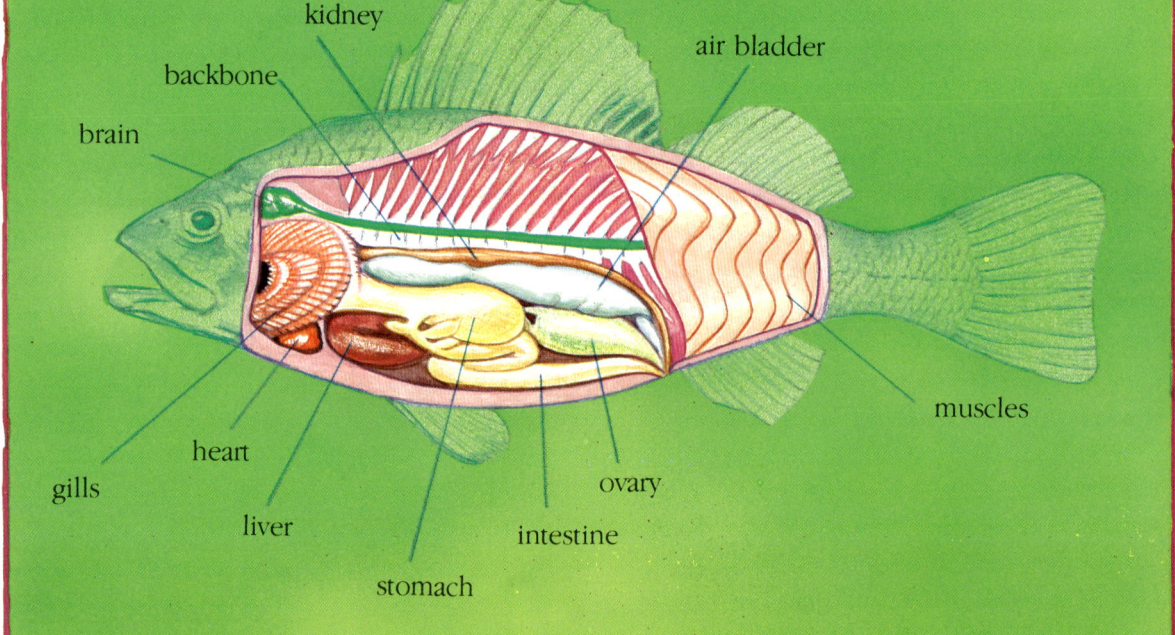
The anatomy of a teleost.

these fish that they slowly but surely began to prevail over the holosts.

And once the latter had practically become extinct, toward the end of the Cretaceous Period, the teleosts, now undisturbed, spread throughout all the seas and fresh waters of the Earth.

THE EXTRAORDINARY EXPANSION OF THE TELEOSTS

The evolutionary drive of the new fish was exceptional. They soon split into many different orders, families and species, probably many more than other classes of vertebrate. Their shape was rather monotonous; typical fish. But some did differ from the general picture. The most curious is the sea horse; with its marsupial pouch, prehensile tail and head out of line with its backbone, it does indeed seem an absurd little creature.

TELEOSTS OUT OF THE WATER

The teleosts did not restrict themselves to invading every type of water available. Like other classes after them, they also tried to leave their habitats. We shall see later how the great expansion of the amphibians first, and then the reptiles and mammals forced some of these to leave the mainland and return to the water. Thus, the teleosts, not content with being masters in the waters, tried to expand in the air and on land. We have the flying fish, for example, that leap out of the water and go on short flights, supporting themselves on very broad front fins spread like wings. Then, the mudskipper (*Periophthalmus*) is a fish that actually manages to live out of water; it can even move around, crawling a little and helping itself along on its front fins, shaped like real limbs. But both of these are awkward creatures, and not much further progress is expected

of them, even in the distant future. We will never see a teleost fly or walk with any confidence. The animal's anatomical organization has evolved so as to give it excellent performance levels as fish, not as birds or mammals. That is to say, our teleosts are too specialized, too closely linked with their environment to undertake such enormous evolutionary changes. The fish that were to cross the boundary between water and land were completely different from the teleosts, as we shall see in the volume on the amphibians.

The basic features of the teleosts are not unlike those we have seen in other fish. Their requirements are the same, so we must not be surprised that animals different in origin and derivation achieve the same solutions.

Flying exocetus

Flying fish, like the exocoetid, make short flights out of the water, gliding along on their broad pectoral fins.

Mudskipper

The mudskipper, common in the rich mangrove regions of Africa and Polynesia, can actually drag itself out of the water and breathe the oxygen in the air.

The Indios of the Amazon River basin make excellent use not only of the meat of the gigantic pirarucu, but also of its scales and bones and even its tongue, so rough that it can be used as a rasp.

Pirarucu

Herrings are typical seafish that live in densely populated shoals.

In the great coral reefs of the Red Sea, the Indian Ocean and the Pacific live some of the most beautifully colored, elegant seafish on earth.

Chaetodon ephippium

Dascyllus

23. THE SHAPE AND COLOR OF THE TELEOSTS

Many teleosts, like herrings, live in great, ever-roaming shoals and are hunted by numerous predators. The few means of defense for the survival of the species lie in their large numbers, which make it reasonably sure that some of the shoal will survive; the large number of eggs laid and, hence, the great reproduction capacity of the female; their ability to evade capture; and camouflage. Generally speaking, these fish have streamlined bodies to allow them to move well in the water. Their colors range from dark shades on the back to progressively lighter shades along the sides and white on the stomach.

CAMOUFLAGE

Its coloring enables the fish to camouflage its body when seen from above, because its dark shape blends well with the seabed; and it uses its white belly to camouflage itself when seen from below, looking like just another reflection on the surface. This second form of concealment does not really work perfectly, but it is helped by the iridescence of the white parts, which reflect light and give off flashes so that the shape of the fish becomes difficult to distinguish. But in spite of all this, they are systematically hunted down. Some do survive, though.

CONSPICUOUS RAIMENT

On the contrary, in the coral reefs of all the Earth's seas, we meet fish with anything but camouflage systems. The most beautiful of them are perhaps the chaetodonts, also called butterfly fish on account of their splendid coloring, rich in contrasts. These fish are so beautiful and tiny that they are kept for decorative purposes in aquariums, where they are admired also for their graceful swimming. They live quite a long time in captivity but, unfortunately, do not reproduce. It is always difficult to analyze what led to certain structures and color schemes, but it is obvious that these kinds of fish cannot have many enemies. It may well be that, the same as birds, their colors convey some message and allow individuals of the same species to identify one another, or to establish dominion over a certain territory, or to attract members of the opposite sex for mating. One thing is sure; these conspicuous costumes did not come about simply by chance.

LIVING IN NESTS

Predatory fish, like the grouper, which live in the darkness of tortuous gorges waiting for prey to swim by, are decidedly dark all over their bodies, if not black. Their shape thus blends in with the shadows or gloom of their dens and cannot be seen by careless victims-to-be.

THE PLEURONECTIDS, OR FLAT FISH

This sub-order of teleosts deserves special mention because of two extraordinary features: their flat bodies and their ability to change color like chameleons. At birth, pleuronectids (side-swimmers), like the sole, the turbot and the flounder, are just like any other baby fish, and they swim around quite happily and normally for a time.

Then, at a certain point of their development, they completely change their habits and the shape of their bodies. They tend to lie on the seabed, on their sides, and their bodies gradually flatten out sideways. And while this change is going on, the eye on the underside — which would be quite useless in that

Grouper

Chaetodon plebejus

The development stages of a flounder, from egg to adult.

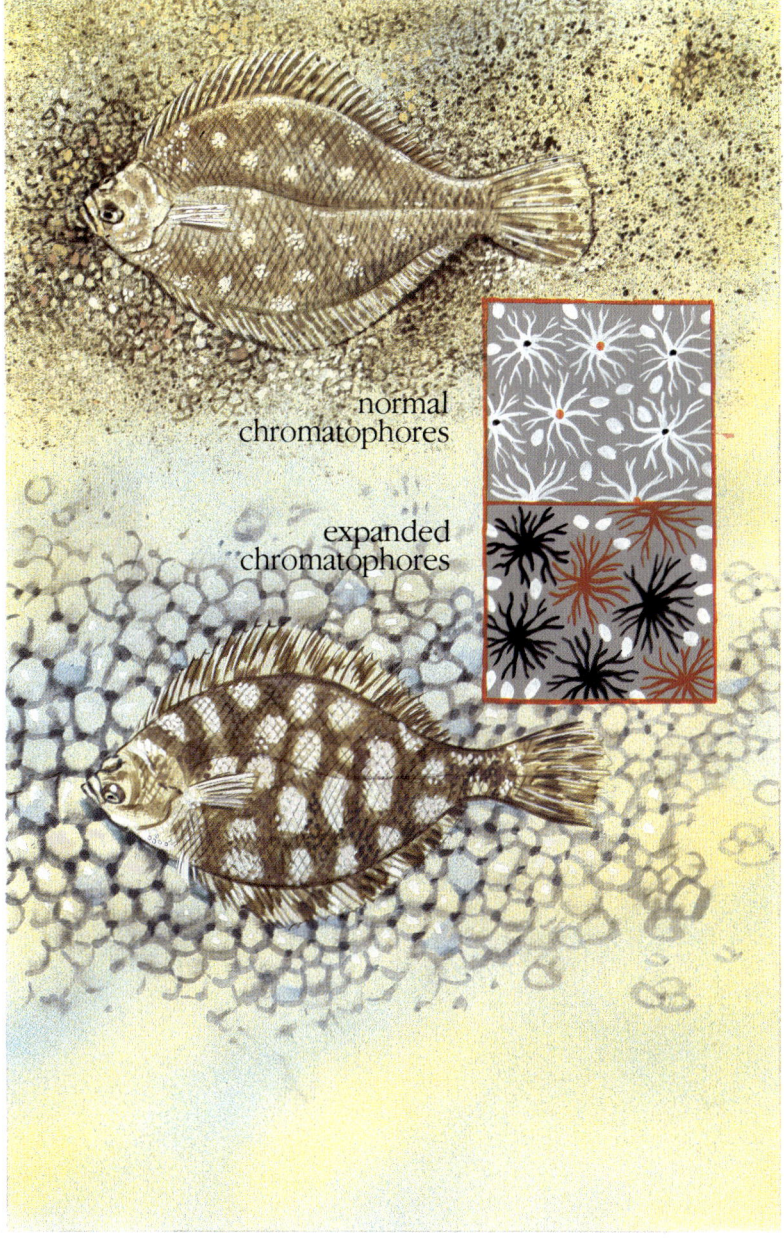

normal chromatophores

expanded chromatophores

position — slowly migrates to the top side, so the animal, flat as it may be, has two apparently normal eyes on one side.

HOW VERTEBRATES CHANGE COLOR

In vertebrates, the coloring is left to special cells called chromatophores (color-bearers), distant relatives of the nerve cells. Chromatophores are cells with many branches and grains of pigment that can move and give or withdraw color. If these grains are gathered around the nucleus, the cell lightens in color, and if all its extensions expand, it acquires the maximum color. All intermediate shades are possible, of course. Many vertebrates are able to change their colors, some rapidly, others more slowly, to harmonize them with the surrounding environment. They do this by a very complicated mechanism that begins with the eye perceiving the color of the surroundings.

Color is produced by five types of chromatophores that take their names from the type of pigment they possess: melanophores are dark brown-black, xanthophores are yellow, erythrophores are red, leucophores are white, and iridocytes provide iridescence. All the various colors are obtained from mixtures of these basic tonalities.

Piranha

The greedy piranhas can strip even bulky victims to the bone in no time at all. In the inset, we see the mouth of a piranha with the terrible triangular teeth just like the predatory shark's.

Some teleosts, like the frog fish, attract their prey with bait provided by their own bodies and suck them in swiftly when they are close enough.

24. THE TELEOSTS: HOW THEY HUNT AND WHY THEY EMIGRATE

We already know of the technique with which the predatory shark attacks its prey and tears off strips of flesh. The teleosts are somewhat more refined and efficient. But some of them still attack their victims directly and shred their flesh ferociously.

THE PIRANHA

Piranha are South American river fish. They are notorious for how swiftly they reduce their victims to mere skeletons. Fording rivers in their territory or simply quenching one's thirst in their waters is an extremely dangerous thing to do. Piranhas are attracted by the odor and literally hurl themselves against the prey and anchor in their razor-sharp teeth. Interestingly enough, their teeth are just like the sharks', flattened triangles with sharp cutting edges. And, again like the sharks, once their teeth have a firm hold on the prey, piranhas thresh their bodies about violently to finish the job and tear off the strip of flesh, which they swallow instantly to be ready for the next inevitable attack. The victim's blood soon spreads through the water, of course, attracting other piranhas that flock in by the hundred, even from great distances, to join in the banquet.

SURPRISE ATTACK

Another technique we can hardly call refined is hunting and direct attack. The pike, for example, uses this method. He lies in wait, concealed amid the vegetation and when the unfortunate victim comes along, he darts out of his hiding-place and snaps at the prey with his sharp teeth. But you need to be an exceptionally fast swimmer to use the surprise factor.

THE SUCTION METHOD

Another fairly common hunting method, which people generally do not know much about, needs no great swimming skills. Indeed, it is normally the heavier, slower, more awkward fish that use it.

The method exploits the breathing mechanism. We have already seen how the fish breathes. It closes its gill chamber, raises its gill sac and opens its mouth to create a negative pressure that sucks in water. The water is then forced through the gills when the fish opens the gill chamber. With a larger gill sac and mouth, which becomes an enormous cavity bristling with teeth, the intake of water can be increased so that the prey is literally sucked right into the awful cavity and to its death. The system works quite well. The grounder uses it, for instance. In fact, it combines the ambush method with instant suction.

FISHER FISH

Other fish have a brilliant method of getting the booty near enough to strike. They use the baiting system. Above the mouth of the frog fish, for instance, there is a long, movable kind of ray with a fleshy, colorful appendage on it, just like a fishing rod.

While the frog fish lies in ambush on the seabed, perfectly still, mouth closed, he agitates his rod frantically. And when some ignorant prey swims along to take the bait, the frog fish opens its mouth violently and in goes the poor wretch. The same system is widely used in the dark depths of the abyss, but the bait this time is a little point of light at the end of the rod.

MIGRATION

One interesting aspect of the teleosts' life that receives a great deal of attention from experts is

Salmon

We all know about the salmon. It lives in the sea and makes enormous journeys, swimming up the torrential rapids of rivers to reach the place where it will reproduce.

Eel

Eels, on the other hand, live in rivers and reproduce in the sea.

Pike

The pike makes a swift, darting attack on its victim.

migration. Most species evolved in an environment they never left, so they can only be found in that kind of environment. Others changed their traditional life style and habits and set off toward different waters. But if the adults' behavior patterns are flexible enough for them to adapt to a new habitat, the requirements of development, which we know little about, still oblige them to return to their places of origin, to emigrate from one environment to another.

So there are massive migrations of fish from the rivers to the seas and vice versa. Let us give two examples.

Salmon used to be freshwater fish, probably living in the shallow, cold waters of the mountains, rich in oxygen. Then, for some reason we are ignorant of, they decided to colonize the seas. So salmon live in the sea, but when they achieve sexual maturity, they return to the rivers they were born in, and swim upstream to the springs, by no means discouraged by the natural obstacles they meet. They then lay their eggs and die. Once the young salmon are well enough developed, they descend the river and return to the sea.

The eel does exactly the opposite. It reproduces in the sea and completes its development in the rivers, migrating the other way round.

One of the most important questions here is just who or what guides the fish. Salmon return to where they were born years before. Eels swim in the sea for thousands of miles. How do they navigate? It seems that salmon use their sense of smell; they remember the odor of their home river, and that guides the creatures. There are vaguer answers as regards the eel. They may take guidance from the stars, or home in on the Earth's magnetic field. Nothing more precise is known.

Certain strong swimmers, like the swordfish, increase the temperature of their muscles to improve their performance.

From their boats, fishermen harpoon the tuna trapped in the "death chamber." The fish thresh about, trying to seek safety, and their frantic movements heat their bodies until the blood becomes "hot and steaming."

25. THE TELEOSTS: A DETAIL OR TWO

Classic zoology textbooks tend to separate vertebrates into two large groups: those that cannot control their body temperature, so that their bodies are always at the same temperature as the environment, as in the case of fish, amphibians and reptiles, and those that keep their body temperature constant, like birds and mammals.

WARM FISH

The great exception is the teleost. Some species are exceptional, tireless swimmers, covering miles a day in search of new hunting grounds. In order to improve their swimming efficiency, they have invented a way of exploiting the heat produced by their moving muscles. When the muscle contracts, some of the energy produced is dissipated in the form of heat, as we realize perfectly well when we sweat after physical work. All fish produce heat when they swim, but it is rapidly dispersed in the water and their bodies stay cold.

But the swordfish and tuna, exceptional swimmers, have a very complex network of vessels that lets the warm venous blood leaving the muscle heat the arterial blood entering it. So the heat they produce while swimming is not lost. On the contrary, it raises the temperature of their muscles. We know that all chemical reactions take place in a given time that is halved if the temperature is above 10 degrees centigrade (50 degrees Fahrenheit). And since muscular contraction is brought about by a complicated series of chemical reactions, increasing the temperature means that swimming is faster and more efficient.

Indeed, the more frantic the swimming, the more the muscles warm up and make contractions become faster too. Scientists have only recently discovered these "warm" fish, which were, however, already well-known to fishermen. Tuna are caught by forcing the school into the so-called "death chamber," an area closed around on all sides and on the bottom by a sturdy net that is gradually brought in closer to the boat so that the fish can be caught. The tuna know that they are prisoners and writhe and thresh about desperately in their attempt to find a way out, but they end up by being harpooned and dragged into the boat. At that moment, say the fishermen, their blood is "hot and steaming." So these teleosts are the only animals in the great group of cold-blooded vertebrates that exploit their own body heat to improve their swimming: an extremely modern, advanced technique. Indeed, we can safely say that these teleosts are far more advanced from the evolutionary point of view than reptiles, the living ones at any rate.

MALES AND FEMALES

Certain sexual aspects of the teleosts are rather difficult to explain. We all know that sex is determined by genetics, by chromosomes. In man, for instance, the two sex chromosomes of the female are identical and are called XX, while the male's sex chromosomes are different and called XY. Whoever possesses one or other of the two pairs becomes a

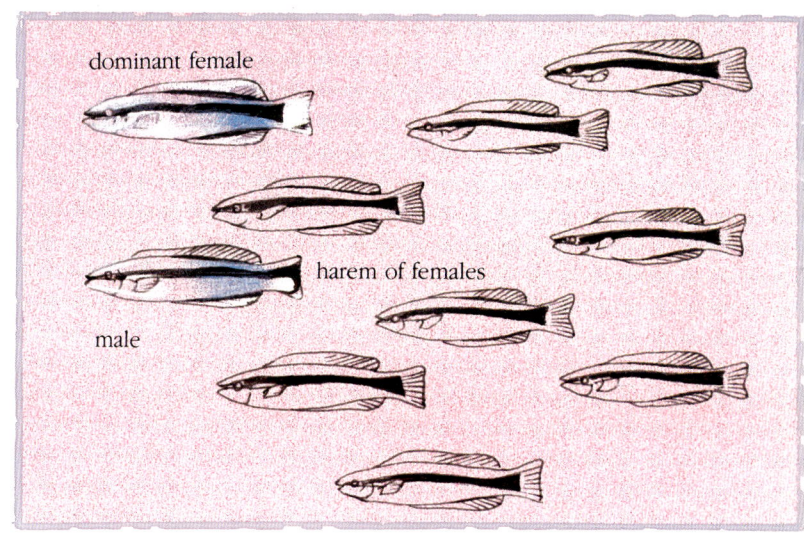

On the death of the male in a shoal of wrasses, comprising one male and a harem of females (above), the dominant female changes sex and takes his place (below).

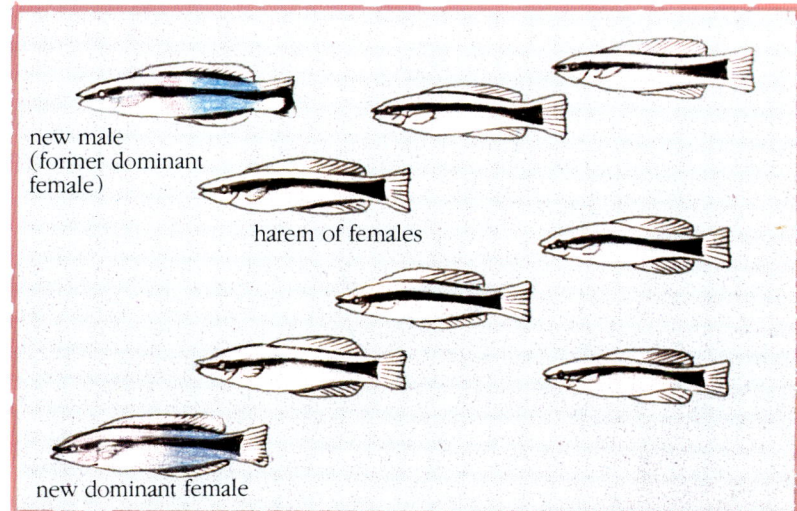

Below: one of the peculiarities of teleosts worth mentioning is the ability of some of them to give off electric shocks. The method is the same as the electric ray's. The electrophore, for instance, can produce a 500-volt current, quite enough to light a lamp and extremely dangerous, even for large mammals.

male or a female once and for all, and the whole organism is endowed with the relatively complex mechanisms for performing the specific functions of sex.

With the teleosts, sex determination is still genetic, but it is not nearly so rigid or schematic. It is far more complex than in mammals.

We have the wrasses, sparids and sea bass, for example. Their behavior is quite peculiar. At birth, the females behave according to their sex; they mate, lay eggs and so on. After a certain number of years, they begin to change sex and take on the features of the male and behave like one. We can try to explain this phenomenon as the consequence of some genetic or endocrinal programming of the organism, but there are other examples that show how fragile the dividing line is between male and female.

The *Labroides dimidiatus*, for instance, lives in small shoals comprising one male and a harem of females. One of the females is dominant, generally the largest one. If, for some reason, the male disappears, the dominant female changes sex and becomes the male while another female takes over as the dominant one, ready to change sex in her turn if the new male, the ex-female, also disappears.

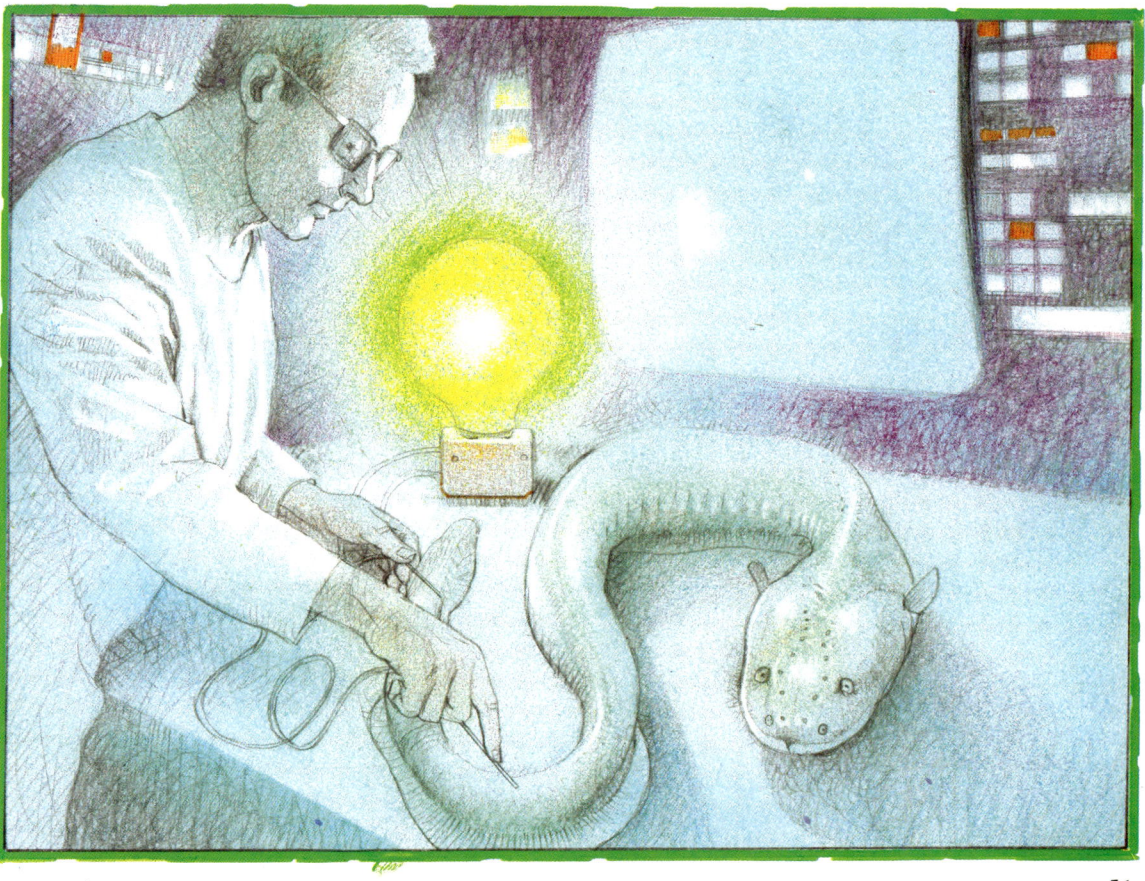

Above: the gonopodium allows the male *Gambusia* to fertilize the eggs inside the female's body.
Below: the *Poecilia reticulata* delivers her young wrapped in a kind of membrane that bursts as they are born.

(1) The courting ceremony of the stickleback. (2) The female enters the nest the male has built and lays the eggs the male will fertilize. (3) Artificial fish shapes made to study the stickleback's behavior.

26. REPRODUCTION AND PARENTAL CARE IN THE TELEOSTS

Most teleosts are oviparous. That is, the eggs are laid outside the female's body and, after fertilization by the male, are left to their fate on the seabed. Mortality rates are very high, of course, both during the development of the eggs and when the fry begin to swim, for they have many enemies. The point is that these fry are an important link in the food chain of sea dwellers, a chain that ranges from microscopic weeds to large hunter fish like the tuna. The survival of the species that reproduces in this way is ensured first by the very large number of eggs one single female lays, often more than a million, and then by the rapid growth and development of the young fish.

Parental care is unknown to these fish and the mating ritual consists of no more than the female laying her eggs on the bottom and the male fertilizing them as he swims over them.

MATING

But mortality rates are drastically reduced when true and proper mating takes place and fertilization occurs inside the female's abdomen. This gives a better chance for the spermatozoon to meet the egg (very uncertain when fertilization takes place outside).

In these teleosts, the male — the *Gambusia*, or topminnow, for example — is easily recognizable, during the mating season, at any rate, because he has a gonopodium, or sex organ, that generally derives from a modification of a ray from the anal fin. After fertilization, the eggs are either laid immediately and abandoned to their fate or kept in nests.

TELEOSTS THAT GIVE BIRTH

There are a number of species, guppies, for instance, that keep the eggs in their abdomens after fertilization, so that the young are able to develop peacefully, safe from all danger. When the fry are ready to swim and find their own food, the mother gives birth. Not many eggs are laid in this species, and

The *Tilapia mossambica* keeps her eggs and young in her mouth.

not many baby fish are born, just a few dozen at each delivery.

THE BEHAVIOR OF THE THREE-SPINED STICKLEBACK

Studying the behavior of fish is a very complicated affair, especially because it is rather difficult to observe them in their natural habitats. In captivity, the creature often loses many of its original behavior patterns. Only the stickleback keeps the same behavior patterns in the aquarium as it had outside, so this fish is widely studied. When the male is ready for reproduction, his belly becomes garishly colorful, red usually, and he becomes particularly vivacious and aggressive.

With great patience, using weeds and a special sticky secretion he produces, he builds a nest and defends it ruthlessly whenever he sees other males approaching. After building the nest, he starts a very carefully determined courting rite and persuades a number of females to enter the nest and lay their eggs, which he methodically fertilizes. And the tireless work of the male is not over yet; he takes care of the eggs, keeping them clean and protecting them from enemies.

It has been asked why the stickleback becomes so aggressive at any trespassing into his area. It has been discovered that it is not the sight of the intruder that triggers off his wrath, because a dummy fish, but not colored red, arouses no reaction in the stickleback. On the other hand, when he sees a roughly oval shape painted red, he immediately becomes enraged. It is the red that upsets him.

AN UNUSUAL NEST: THE MOUTH

The *Tilapia* is an African freshwater fish that is being bred worldwide because of its tasty meat and its capacity to adapt and reproduce without difficulty. This fish invented a truly singular way of protecting eggs and fry. Fertilization occurs outside the female's body, but she sucks the fertilized eggs into her mouth and keeps them there while they develop. At birth, the fry leave their mother's mouth but do not stray too far. At the slightest sign of danger, there they are, straight back in the safety of their mother's mouth. Not for nothing is this fish also called "mouthbreeder."

MONSTER FISH

One of the manifestations of the decorative tastes of the Japanese and their admirable age-old patience is that they have been able to dramatically modify the common goldfish, the *Carassius auratus*. The results are anatomically absurd shapes, monstrous in the eyes of a naturalist, but pleasant for the Japanese. Not much is known about how these monsters are formed, but they are based on mutations that occur not during development, because the fish are normal at birth, but during growth.

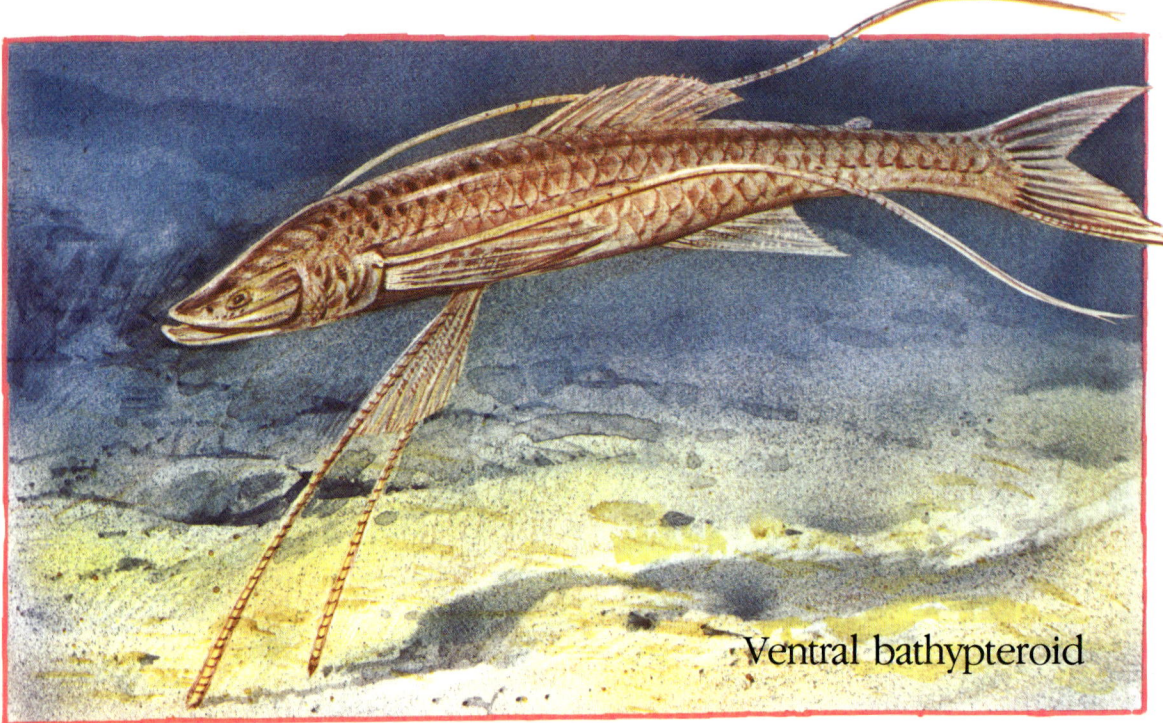

The ventral bathypteroid is a fish of the depths that lives on the seabed.

27. THE TELEOSTS OF THE ABYSS

HOW TO MAKE SURE YOU NEVER LOSE YOUR HUSBAND

In the darkness of the abyss, encounters between the two sexes are indeed casual. Light signals sometimes help them meet, but the problem is always the same. The *Photocorynus*, a teleost, solves the problem in its own inimitable way. The male is much smaller than the female. When by chance he does meet one, he literally welds his body to her. His tissues enter into symbiosis, or organic connection, with hers, and he feeds off her blood.

The welding is permanent and ceases only with the death of the female, who could never wish for a more faithful husband.

At the present time, we know more about the surface of the moon than about the depths of our seas. There are enormous difficulties involved in exploring this part of the world: high pressures, wide, open spaces, different currents at different depths, weather conditions. What we do know comes from the few fishing expeditions that have used special nets and specially fitted out ships to catch deep-sea fish, and the even rarer descents in bathyscaphes with men and TV cameras on board. The knowledge we have of our ocean depths is definitely a partial knowledge. There must still be astonishing discoveries yet to be made here.

IN THE UNKNOWN DEPTHS

We do know that, in the everlasting darkness of the marine abyss, life is represented by all the great zoological groups, vertebrates too. And occasional guests often drop into these depths to seek food. We know that many sharks often swim down deeper than 5,000 feet. And we know that certain cetaceans, including the courageous sperm whale, perhaps, can go down even deeper. Exhaustive studies have been made of this whale, because enormous, apparently insoluble problems crop up when a mammal dives from the surface to the absolute depths in no more than 20 to 30 minutes. Just think of embolism, when a blood clot blocks a vessel, the great problem for divers when they go just 20 or 30 yards below the surface.

THE PERMANENT RESIDENTS OF THE ABYSS

Resident animal life in the marine abyss arouses a good deal of interest. First and foremost, there can only be fish that hunt other fish in this environment. At this depth, the food chain cannot begin with plant life followed by plant eaters and so on. It must begin immediately with animals that devour others. But if that were all, life in the abyss would soon cease altogether; few animals would survive. What we have to realize is that, above the abyss, there are thousands of fathoms of sea where other fish live, and they fight, grow old and die. Hence a veritable shower of organic substances, débris, scraps of fish and dead fish pours without end down into the abyss to sustain life at such great depths. Most of the great groups of invertebrates are there, from the annelids to the mollusks and arthropods, and their forms are not very different from the ones that live in shallower waters. And here the teleosts become extremely impressive predators. They have huge, monstrous mouths, as in the *Eupharynx* (good-throat), which can capture and gobble up fish much larger than itself. And sometimes the silhouette of the swallowed prey can be seen in the predator's stomach, as in the *Saccopharynx* (bag-throat). Teeth are often very long and sharp, out of proportion almost.

Then, other species have luminous organs that sometimes help to make the surroundings a little lighter but are more often used to send signals from one individual of the same species to another, or to help one sex to recognize the other, or to attract prey, as we have already seen.

Opposite page: in the darkness of the marine abyss, permanent residents have taken on fantastic shapes, exaggerating the features of the predator.

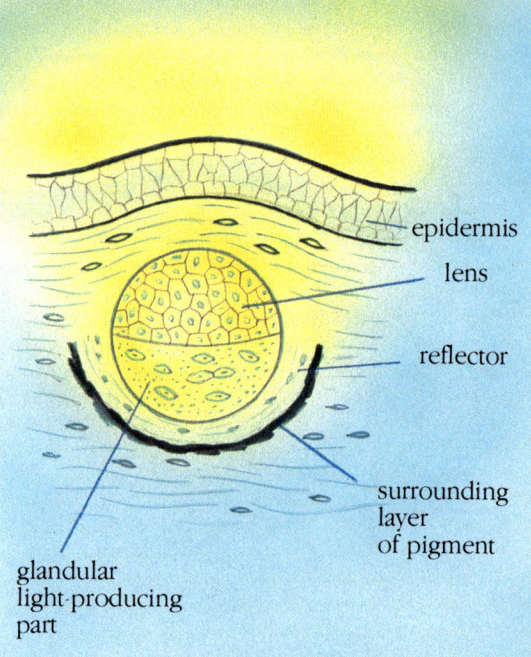

GLANDS THAT PRODUCE LIGHT

Light organs derive from the cells of glands in which a chemical reaction produces light. Then, there are small accessories, such as crystals that act like reflectors, and cells that act like lenses and concentrate light in beams. Other organs take their light from bacteria that can trigger off the necessary chemical reaction. In this case, the only function of the gland is to feed these bacteria.

28. SARCOPTERYGII: THE FISH THAT ALSO BREATHE AIR

Opposite page: the *Sarcopterygii* evolved and adapted in a marshy environment, with warm waters poor in oxygen. Some of them swam or crawled on the bottom (below) while others rose to the surface to breathe ordinary air (above).

The acanthodians generated both the *Actinopterygii* and the *Sarcopterygii*, fish that breathed ordinary air.

We have already seen how the Devonian Period was the great period of fish. Nearly 400 million years ago, in fresh waters above all, fish with bony skeletons appeared and expanded. The ancestors of them all were the acanthodians, small fish covered with spines.

Two great phyla originated from them and proceeded with their evolution along quite separate paths, achieving completely different anatomical and functional organizations linked with the environments that had selected and perfected them.

The path of the *Actinopterygii*, which we have seen, was followed by typical fish, completely bound and adapted to water. All the features of their anatomies were shaped for life in more or less deep waters.

The other path, however, that of the *Sarcopterygii*, was followed by bony fish that adapted to life on the boundary between water and air.

IN THE WARM WATERS OF THE DEVONIAN

A reconstruction of environments as they were in the Devonian Period, based mainly on studies of the plant life and geological features of the period, leads us to believe that land above water was mostly marshland, with slow, shallow rivers flowing down from the few mountains. The climate must have been permanently warm or hot. These climatic conditions persisted for more than 120 million years and deeply affected the evolution of the *Sarcopterygii* and the origin of the amphibians.

So, because the fresh waters of the period were so shallow and warm, we may conclude that they contained little oxygen. Like any other gas, this important molecule dissolves in water in greater quantities if the temperature is low. Turbulent and bubbling, mountain streams do in fact contain a good deal of atmospheric oxygen. If water is hot, on the other hand, not only does the atmospheric oxygen not dissolve in it, but the oxygen its plant life produces by photosynthesis immediately becomes gaseous, rises to the surface and is dispersed.

Fish normally breathe through their gills and take their oxygen from the water, so living in warm water is difficult for them, if not impossible. If the first *Actinopterygii* were freshwater fish, the fish that derived from them preferred the deeper, cooler, oxygen-rich seas. Only later, many millions of years later, when the climate of the Earth changed and the fresh waters became deeper and cooler, did the *Actinopterygii* return to the places of their origin. But in that period of time, from the Devonian right through the Carboniferous Periods, the ecological niche of fresh waters poor in oxygen was occupied by fish that could also breathe ordinary air.

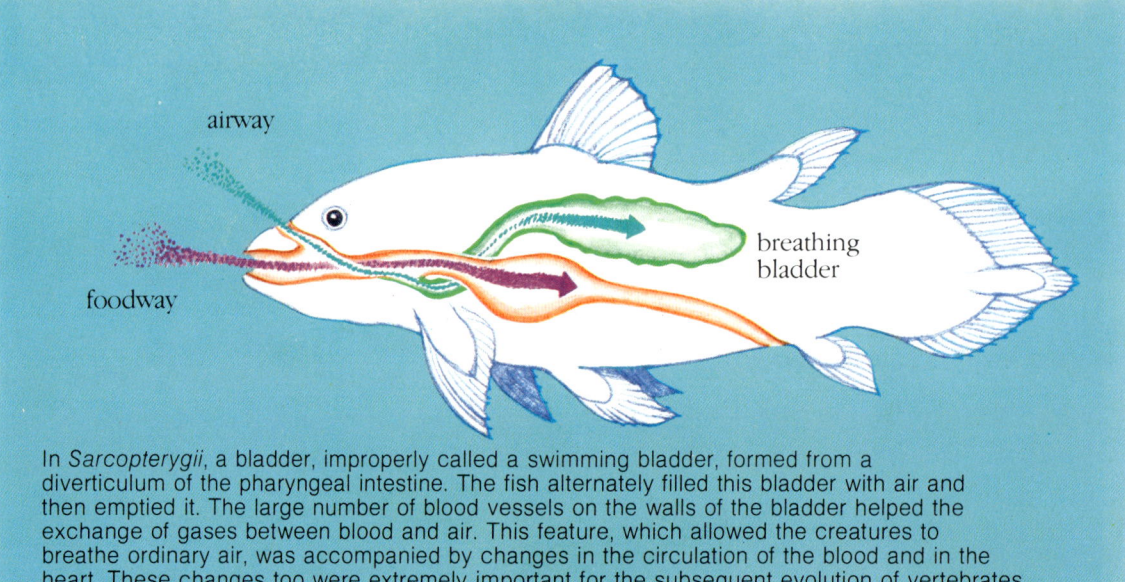

In *Sarcopterygii*, a bladder, improperly called a swimming bladder, formed from a diverticulum of the pharyngeal intestine. The fish alternately filled this bladder with air and then emptied it. The large number of blood vessels on the walls of the bladder helped the exchange of gases between blood and air. This feature, which allowed the creatures to breathe ordinary air, was accompanied by changes in the circulation of the blood and in the heart. These changes too were extremely important for the subsequent evolution of vertebrates.

FISH THAT BREATHE AIR AND BEGIN TO WALK

The *Sarcopterygii* were the first great group of fish to adapt to this difficult environment.

The long period of time allowed the evolution of forms suited to living in water and using ordinary air. They needed, and got, a special organ for breathing in this environment, but that was not all; the shallow water often meant that fish had to drag themselves across muddy ground. To help them do this, the classic fins used for swimming became stumpier, more robust, able to bear the weight of the animal's body and move it along. Thus, for the first time in vertebrates, a new way of moving appeared. These were not yet the legs of the amphibians and other land-bound vertebrates, but they were the evolutionary precursors needed for movement on limbs. The two important features of the sarcopterygii — the new breathing organ and the stubby fins — which evolved as they adapted to their special environment, were the foundations for the evolution of the amphibians, a highly important stage in the history of vertebrates. It is a classic case of pre-adaptation; an anatomical feature that evolved for a specific purpose was then perfected by subsequent evolution and eventually led to spectacular, unforeseeable results.